Cord Cutting

by Paul McFedries

for
dummies
A Wiley Brand

Cord Cutting For Dummies®

Published by **John Wiley & Sons, Inc.**, 111 River Street, Hoboken, NJ 07030-5774, www.wiley.com

For general information on our other products and services, please contact our Customer Care Department within the U.S. at 877-762-2974, outside the U.S. at 317-572-3993, or fax 317-572-4002. For technical support, please visit https://hu.b.wiley.com/community/support/dummies.

Wiley publishes in a variety of print and electronic formats and by print-on-demand. Some material included with standard print versions of this book may not be included in e-books or in print-on-demand. If this book refers to media such as a CD or DVD that is not included in the version you purchased, you may download this material at http://booksupport.wiley.com. For more information about Wiley products, visit www.wiley.com.

Library of Congress Control Number: 2021936887

ISBN: 978-1-119-80093-4; ISBN (ePDF): 978-1-119-80097-2; ISBN (ePub): 978-1-119-80098-9

Manufactured in the United States of America

SKY10026499_042621

Contents at a Glance

Table of Contents

PART 4: THE PART OF TENS

CHAPTER 19: Ten Ways to Save Money in a Cord-Free World

CHAPTER 20: Ten Tips for Troubleshooting Streaming Woes

Introduction

William the Conqueror, it is said, began by eating a mouthful of English sand.

— SALMAN RUSHDIE

A s I write this, more than 40 million people in the United States alone have cut the cord and banished the cable company from their lives. What we have here, beyond any doubt, is a genuine phenomenon. We're witnessing a kind of cord-cutting frenzy as people of all ages, all regions, and all walks of life thumb their noses at traditional cable and opt, instead, for the freedom of a cable-less lifestyle.

If you're looking to join this movement, I have some good news right off the bat: Unlike William the Conqueror, you don't need to eat a mouthful of sand to get started. Ah, I hear you ask, but where *do* I start? If you have even a passing familiarity with watching TV without cable, you know that it's a sprawling, labyrinthine, and constantly changing topic. So many shows! So many channels! So many services! So many devices! Where, indeed, do you start?

I'm glad you asked, because that's exactly where *Cord Cutting For Dummies* comes in. I've traveled the entire cord-cutting landscape from east to west and from north to south. I've connected the devices, subscribed to the services, and installed the apps. The result is the book you're holding (physically or virtually).

About This Book

Cord Cutting For Dummies shows you not only how to get the cable company out of your life but also what to do after that. This book takes you on a tour of all the main (and a few minor) ways to watch TV without cable.

In Part 1, you learn lots of good reasons why you should cut the cord (plus a few sensible reasons why going cordless might *not* be a good idea for you) and you get a step-by-step plan for going cord-free.

In Part 2, you get your post-cord life off to a free and easy start by learning all about over-the-air TV, where you get high-quality channels free (yep, that's right: *free*). You learn how over-the-air TV works, what equipment you need, and how to set everything up.

In Part 3, you dive into streaming services such as Netflix, Amazon Prime Video, and Hulu. You learn about streaming media players and smart TVs, and how to get your Internet access ready for streaming. You then check out a bunch of free and paid streaming services.

Finally, in Part 4, you learn ten ways to save money after you've cut the cord and ten tips for troubleshooting streaming problems.

The chapters present their info and techniques in readily digestible, bite-size chunks, so you can certainly graze your way through this book.

Foolish Assumptions

Cord Cutting For Dummies is for people who are new (or relatively new) to cord cutting. This doesn't mean, however, that the book is suitable for everyone. I've made a few assumptions about what is necessary if you want to flourish in a cable-free world. Here they are:

>> You know how to connect devices to your TV.

>> You have an Internet connection and a Wi-Fi network.

>> You can use a web browser to navigate to a particular website given that site's address.

>> You know the basics of launching and using mobile device apps.

That's about it, really. Cord cutting is a big topic, but it's not one that requires a huge amount of technical literacy. And what literacy you *do* need I explain as we go along.

Icons Used in This Book

Like other books in the *Dummies* series, this book uses icons, or little margin pictures, to flag info.

REMEMBER

This icon marks text that contains info that's useful or important enough that you'd do well to store the text somewhere safe in your memory for later recall.

TECHNICAL STUFF

This icon extra information that either is a bit on the advanced side or goes into heroic, often obscure detail about the topic at hand. Do you need to read it? Not at all. Does that make the text a waste of page real estate? I don't think so, because the information is useful for folks interested in delving into the minutiae of cord cutting. If that's not you, ignore away.

TIP

This icon marks text that contains a shortcut or an easier way to do things, which I hope will make your life — or, at least, the data analysis portion of your life — more efficient.

WARNING

This icon marks text that contains a friendly but insistent reminder to avoid doing something. You have been warned.

Beyond the Book

To locate this book's cheat sheet, go to www.dummies.com and search for *Cord Cutting For Dummies*. See the cheat sheet for some recommended streaming media services for kids' programming, sports, and news.

Where to Go from Here

This book consists of a couple of hundred pages. Do I expect you to read every word on every page? Yes, I do. Just kidding! No, of course I don't. Entire sections — heck, maybe even entire chapters — might contain information that's not relevant to you.

In *Alice's Adventures in Wonderland*, the King of Hearts tells Alice that she should, "Begin at the beginning and go on till you come to the end: then stop." But you don't have to follow his advice here.

However, if you're just getting started down the cord-cutting road — particularly if you're not sure you even *want* to cut the cord — no problem: I'm here to help. To get your cord-cutting education off to a solid start, I highly recommend that you start with Chapter 1 to find out if you really do want to go cable-free. If you do, continue with Chapter 2 to find out how it's done. From there, you can travel to the post-cable territory of Chapters 3 and beyond.

If you have some experience with cord cutting or you have a special interest or question, see the table of contents or the index to find out where I cover that topic, and then turn to that page.

Either way, happy cord cutting!

1

Some Cord-Cutting Basics

IN THIS PART . . .

Discover some excellent reasons why you should — or maybe shouldn't — cut the cable TV cord.

Follow a step-by-step guide to cutting the cord.

IN THIS CHAPTER

» **Having fun dissing the cable company**

» **Seeing if cutting the cord is right for you**

» **Understanding the benefits of cutting the cord**

» **Touring a world without cable**

Chapter **1**

Why Cut the Cord?

Y ou can get a TV signal into your home these days in many ways, but most methods involve running some kind of cord into your home and then into a device. That device might be a TV, a set-top box, or even a cable modem.

So far, so obvious. My point here is to bring your attention to the cord itself, which carries this book's symbolic load. Why? Because a new and growing legion of people are using their cable TV cord as a symbol for protest.

Who are these people? The *cord shavers* (also called *cord trimmers*) take steps to reduce their cable TV bill. The *cord avoiders* look for online alternatives to paying for cable TV offerings. Next are the *cord haters,* who really dislike paying for cable TV. All these people look on in envy at the *cord nevers,* people who have never had a cable TV account.

The cord shavers, cord avoiders, and cord haters can't be cord nevers, but they can certainly become *cord cutters.* Those are people who metaphorically snip their cable TV cord by non-metaphorically canceling their cable TV subscription and looking for televised entertainment elsewhere.

If you find your blood pressure getting dangerously high each time you pay your cable TV bill, you might be ready to become a cord cutter. To make sure, this chapter talks about why you might want to go the cord-cutting route (and a little bit about why you might not).

What's So Bad about Cable? (Let Me Count the Ways)

Every year, various media organizations publish articles with titles along the lines of "The Ten Most Hated Companies" or "The Twenty Worst Companies." A wide variety of industries is represented, from airlines to social media companies to banking institutions. The lists change year to year, but you can always count on at least one or more cable companies making the list.

Let's face it: Many of us *really* dislike our cable provider. What about you? How do you feel about the company that brings cable TV into your home?

Okay, you're reading a book about cord cutting, so I have to assume that you're at least peeved at your cable company. Or maybe a bit miffed. But however you feel, you might need coaxing before you go any further. Sure, I understand: Cutting the cord is a big step. To help you make your decision, this section details ten solid reasons why you might want to cancel your cable subscription and join the ranks of cord cutters.

Cable TV is expensive

Aside from essential utilities (heat, electricity, water, phone) and expenses such as groceries and a car payment, most of your regular monthly payments probably don't amount to that much money. Five dollars here, ten dollars there, twenty dollars somewhere else. Then your monthly cable bill comes due and, by contrast, it's probably a whopper: Depending on your channels, packages, equipment, and bundles, you can easily be paying a hundred, two hundred, even three hundred dollars or more — a *month!*

As much as you might enjoy the cable company's offerings, that cable bill qualifies as an extravagance. Now, as you soon see, money isn't the only reason to cut yourself free from cable, but for most would-be cord cutters, it's the reason that gets people thinking there has to be a better (and cheaper) way.

You still get bombarded by commercials

You pay your budget-busting cable bill and now you're stressed and angry. What's the antidote? You know: Watch a little TV. But when you turn on the set, chances are the first thing you see is a commercial. Then another one. And another. Sure, you're a savvy TV pro, so you know where to find the mute button on your remote.

But still: Doesn't it rankle? You pay a queen's ransom for (apparently) the privilege of watching TV, only to be subjected to endless come-ons for hemorrhoid remedies and car insurance. That's just wrong.

You probably watch only a teensy fraction of what you pay for

The Bruce Springsteen song "57 Channels (and Nothin' On)" was released way back in 1992, but it's still relevant today. Except now it's closer to 557 channels. However many channels come with your cable package, it's a safe bet that you find a depressingly vast majority of them unwatchable so they are therefore unwatched.

Sure, you have your favorite stations, but how many do you watch regularly? A dozen? Maybe a couple dozen? That still leaves hundreds of channels gathering dust. Even worse: You're paying for those dust-covered channels. Why would anyone do that?

Bundles are (usually) bad

The standard way to save money when it comes to the cable company is to invest in a *bundle*: a collection of cable company services that includes not only cable TV but also Internet access, a home phone, a cellular plan, or some combination of these and other offerings. Instead of paying for each service by itself, you bundle them for a substantial discount.

That makes sense, but there's a fly in the bundle ointment: Almost always, at least one of the bundled services will be *terrible*. It might have cheap — or even used! — equipment, spotty service, minimal features, or (all too often) all of the above. Even though you save money with a second-rate service, you shouldn't have to live like that.

Long-term (read: loyal) customers pay more, not less

If you're a regular customer at your local coffee shop, every now and then the manager might slip you a free coffee or muffin. If you buy lots of clothes at a local independent retailer, the owner might give you a discount on a larger purchase. What these small businesses understand is the value of customer loyalty: It pays in the long term to keep regular customers happy.

Your cable company understands this, too, right? Hah, don't make me laugh! In fact, your cable provider probably does the opposite. That is, they probably offer discounted subscriptions to new customers, while charging substantially higher prices to long-term subscribers. It's madness, but welcome to the wacky world of the cable company!

REMEMBER

If you plan on sticking with your cable subscription, at least for a while, you can almost certainly negotiate a lower price. Call the cable company, complain about how high your monthly bill has become, and then threaten to either switch to another provider (assuming there is one where you live) or cancel your subscription. It might take some perseverance on your part and a session with someone in the Customer Retention department, but you'll get there.

You're getting nickeled-and-dimed

If you've ever been brave (or foolhardy) enough to examine the details of your cable bill, chances are you came away even angrier than you were going in. It's not just the sheer size of the final total that stresses everyone out, but the long list of incidental and hidden fees and surcharges that are standard-issue line items in every cable bill. These fees go by various names:

>> Network access fee

>> Broadcast TV fee

>> HD technology fee

>> Regional sports surcharge

>> Terminal fee

>> Franchise fee

The list goes on and on and includes lots of regulatory fees mandated by the government, which the cable company is happy to pass on to you as so-called *pass-through fees*. These fees and surcharges easily cost you hundreds of dollars a year. Boo!

You're dealing with a near-monopoly

Mom-and-pop shops in the cable business don't exist because cable requires huge infrastructure investments. That's fine, but with recent consolidations in the industry, the gargantuan just keep getting gargantuan-er. The result is that even large markets have few options for cable TV providers, and small markets might have just a single company "vying" for their business.

This lack of competition is bad news for you. It keeps cable prices high, encourages cable companies to charge — and increase — hidden fees (as I describe in the preceding section), and gives cable behemoths zero incentive to provide decent customer service and technical support (travesties that I talk about in the next two sections).

Customer service is pretty much non-existent

Earlier I mention that cable companies always show up in lists of the worst or most hated companies. As this section shows, there are lots of reason why that's so. However, one of main complaints you see in surveys of customer (dis)satisfaction is terrible customer service.

You know what I'm talking about, right? Does anyone ever look forward to calling the cable company? Having such a call on your to-do list is likely to elicit feelings of dread and anxiety because the poor souls who work in a cable company's customer service department aren't allowed to be human beings. Instead, they're browbeaten into giving rote answers that never deviate from a management-approved script.

And if you get even a little frustrated or upset at the runaround you're getting, there's an excellent chance the rep will put you on hold forever and then simply disconnect the call!

TIP

You can see for yourself how bad cable company customer service is by taking a look at the American Customer Satisfaction Index for Subscription Television Services at www.theacsi.org/index. php?option=com_content&view=article&id=147&catid=&Item id=212&i=Subscription+Television+Service.

Technical support is a pain in the you-know-what

Calling the cable company's technical support department is no better an experience than the customer service nightmare I moaned about in the preceding section. First, you have to wait on hold for a very long time. Second, the "technician" (note the sarcastic quotes there) will ask about your problem, and then spend a *very* long time going through an infuriatingly banal and wrong-headed flowchart-slash-script in an attempt to find a "solution" (more sarcastic quotes).

That approach *never* works, so now the rep will book you an appointment with an actual technician. Alas, the next available appointment is in two weeks (if you're lucky) and, yes, you'll have to take half a day off work. Oh, and it will set you back $50, $75, or even $100 just for the technician to show up.

Those darned contracts!

As I mention, you can often negotiate a lower cable bill by putting together a bundle of services, packages of content, or both. But there's usually a catch, actually *two* catches: You have to sign a contract (usually for two years) and the discount applies only for the first year! So you're stuck paying a higher price for the rest of the contract, unless you agree to pay an exorbitant termination fee to opt out. Grrr.

WARNING

After signing the contract, you'll receive a confirmation, usually by email. Double-check — no *triple*-check — the order to make sure you're getting what you asked for and what was promised to you. Cable company sales reps work on commission and will often simply modify orders — while betting that you won't notice — if doing so benefits them.

Some Reasons Why Cutting the Cord Might Not Be for You

This book is about cutting the cord, but I might as well admit early on that there's no perfect solution to going cable-free. My thesis here is that, for most people, saying goodbye to the cable company is a net win. However, a few aspects of cutting the cord fall on the "cons" side of any "pros versus cons" analysis, and one or more of those could be a deal-breaker for you. Let's see.

Your savings might be less than you hoped

Everyone goes into the cord-cutting adventure with big dreams of saving a ton of money every month. And those savings *are* possible, especially if you embrace free and almost-free services. However, most of the good content sits on the other side of a paid subscription.

You can still save lots of cash if you're prudent with your subscriptions. Unfortunately, many families find that they keep adding new services (particularly for popular content such as Disney, HBO, news, and live sports) and their monthly TV-watching costs rise accordingly.

You can use lots of tricks and techniques to save money after you cut the cord. I talk about a bunch of these in Chapter 10.

You might still have to deal with channel bundles

Most of us hate channel bundles because to subscribe to the one channel you want, you also have to get a fistful of channels that you wouldn't force your worst enemy to watch. So now it feels like you're paying the bundle fee for just a single channel. Cue the steam coming out of your ears.

Bundles aren't an issue with subscriptions such as Netflix, where one price gets you access to everything on the service. Unfortunately, far too many streaming services embrace the bundle model and surround premium content (such as HBO) with dreck.

You might still see commercials

For lots of would-be cord cutters, the real incentive is not cost savings but a commercial-free viewing experience. The good news is that most streaming services are on board with the commonsense notion that you shouldn't see commercials if you pay a subscription for the service. Sweet bliss!

However, some free streaming services *do* show commercials, because they have to pay their bills somehow. And, after all, seeing the odd commercial is a small price to pay for a free service.

Alas, just because a streaming service doesn't currently show commercials, it doesn't follow that the service will always be ad-free. For example, Netflix, which is currently commercial-free, has run tests in which they show commercials between TV show episodes.

TV watching will become more complex for you

You can bad-mouth cable companies all you want (and I know you do), but they do have one genuinely good feature: simplicity. Sure, you pay a ton of money each month, but in exchange you get all your channels and apps and more in a single package with a single interface.

Once you cut the cord, that simplicity will probably become a thing of the past. I say "probably" because it *is* possible to create simple cord-free experiences where, say, you watch only live TV or you subscribe to only a single streaming service. But you're more likely to end up with multiple subscriptions on multiple services. That means paying multiple bills, configuring multiple accounts, and learning multiple app interfaces. And you also run smack into a very modern problem: trying to remember which streaming service offers which content!

I wish I could tell you that the cord-free experience is getting simpler, but the opposite is happening. Media companies are falling all over themselves to launch their own streaming services. Whereas a few years ago you might have been able to count the number of streaming services using the fingers of one hand, the way things are going you'll soon need the fingers of every member of your extended family.

You'll use way more Internet bandwidth — and perhaps pay for the privilege

Streaming media comes to you via the Internet, where it's distributed through Wi-Fi to your various devices. You don't normally give it a second thought, but perhaps you should. Why? Because media streams such as movies, TV shows, and music stuff huge amounts of data into the pipe that delivers the Internet to your home. If your Internet service provider (ISP) puts a limit on your monthly bandwidth, blowing through that cap because you binge-watched *Better Call Saul* may cost you a ton of money.

REMEMBER

If you have an ISP plan that offers unlimited bandwidth (lucky you!), you don't have to worry about any of this because there's no ceiling to go through.

Your monthly Internet bill will probably go up

When most people are budgeting for a cord-free lifestyle, they usually compare their total cable bill with what it might cost for a few streaming services. That's a legit comparison if all you get from the cable company is cable TV. However, if you get multiple products — such as cable TV, Internet, and phone service — the comparison falls apart because the Internet portion of that bill is almost certainly discounted.

When you tell the cable company that you no longer want their stinkin' cable TV service, the first thing they'll tell you is that your monthly Internet bill will go up because you no longer have a proper bundle of services.

The quality of the streaming video might be poor

I talk in Chapter 7 about the Internet connection you need to support a cord-free life. For now, you just need to know that to be able to watch streaming media — particularly streaming video — you must have fast Internet access. How fast? The necessary download speed — measured in Mbps (megabits per second) — depends on the video quality you're streaming.

If your Internet download speed isn't fast enough, your streaming experience will be poor: slow starts, playback pauses and stutters, and overall lousy picture quality.

You might have to wait a long time to watch new shows

When a new cable show (or a new season of an existing cable show) is released, cable providers get first crack at broadcasting it. When you banish the cable company from your life, you also lose the privilege of being first in line to watch this new content.

Sure, there's a decent chance that the new episodes will eventually end up on some other service, such as Netflix. But *eventually* is the operative word and often means up to a year after the show's release.

You might never see some new shows

Some folks are calling this the Golden Age of television because so much great content is being created. Think of shows like *The Handmaid's Tale*, *The Queen's Gambit*, *The Morning Show*, and *The Marvelous Mrs. Maisel*. Besides starting with the word *The*, each of these shows is original programming created by a streaming service: Hulu, Netflix, Apple TV+, and Amazon Prime Video, respectively. It's great that these services are pouring so much money into creating amazing television, but the downside is that the only way to see each of these shows is to have a subscription to the service that created it.

Is it possible that some of these shows might appear down the road on another service? Maybe, but I wouldn't count on it.

The Benefits of Severing Your Connection to the Cable Company

Given all the reasons listed near the beginning of this chapter as to why the cable company is so awful, clearly the main advantage to cutting the cord is never having to deal with your cable provider again! However, although getting Big Cable out of your life forever is a huge benefit, it's not the only one you get when you snip that cord. Let me take you through a few more.

You save money

Carving a sizable chunk off your monthly television-watching bill is the most common reason most people dream about cutting the cord. Sure, earlier I talked about how you might not save as much as you'd like, but how much you save depends on you. If you're happy to just "Netflix and chill" every night, you're going to save a ton of money each month. If you're a sports nut or news junkie, or if your TV tastes run towards premium channels such as HBO, you'll pay extra for the privilege.

Don't get me wrong: Cable subscriptions are so expensive these days that almost everyone ends up paying less each month when they cut the cord.

You'll probably be happier in the long run

Cable TV has what smarty-pants economists call *low perceived value*, which means you don't feel like you get your money's worth from your cable subscription. That is, although you pay a ton of money each month, you enjoy only a few shows, are indifferent to (or downright dislike) many more, and ignore the majority of what's available. That almost-no-bang-for-your-buck experience is depressing.

By contrast, surveys of cord cutters routinely show great satisfaction, which comes from having a *high perceived value* of the new lifestyle. To be accurate, the cord cutters who are happiest with making the change are those who've kept things simple by subscribing to only a few streaming services. The combination of saving money, having a simple setup, and being satisfied with what the services offer is the ticket to cord-cutting nirvana.

You unshackle your TV

A big problem with cable TV is the cable itself. Sure, lots of TV comes to your TV via Wi-Fi these days, but many people still have a cable outlet on one wall and a coaxial cable running from that outlet to a set-top box, which then connects to your TV. This setup is fine as long as you're okay with having your TV (and therefore the rest of your entertainment center) relatively close to the cable outlet.

However, what if one day you decide that your TV-watching experience would be much better if you could move your TV to the opposite side of the room? Well, sure, you *could* do that, but it means buying a really long coaxial cable and stringing it along the base of your walls to the new TV location. That's ugly with a capital "Ugh."

And if you decide your TV should be on a different floor? Ah, now you're looking at the expense of bringing in a cable technician to move the outlet.

Cutting the cord, by contrast, means literally doing away with that freedom-restricting coaxial cable. With your content now coming in over the Internet and broadcast to your smart TV or your streaming device via Wi-Fi, moving your TV to the perfect location is easy.

REMEMBER

The big assumption behind this benefit is that you're *not* getting at least part of your TV fix using an over-the-air antenna, which still requires a coaxial cable connection to your TV.

You unshackle yourself

After you go cord-free, that freedom extends where you watch TV. You're no longer required to plop yourself down, potato-like, on the couch in front of your TV set. Instead, because every streaming service and device offers an app not only for configuration but also for viewing content, you can use your favorite tablet or smartphone to watch shows anywhere you want.

In the living room? Of course. In the bedroom or the kitchen? Sure. In the den? Perfect. In the bathroom? Um, your call.

You're in control

So much of the cable TV experience feels like coercion. The basic or standard package has ten stations you watch regularly, and fifty you didn't even know existed. A specialty package has one or two channels you want, and eight or nine channels that do nothing for you. You need a set-top box, which the cable company is happy to lease to you forever at ten or twenty bucks a month.

When you throw down your scissors after cutting the cord, that lightness you feel is the lifting of these and similar cable company burdens. Now *you* are in control, deciding which channels or services and equipment you want. Ah, that's better.

Surveying the Cord-Free Landscape

What can you expect to find in a world where the cable company is a distant memory? Answering that question is what the rest of this book is about. To give you a feel for what's in store, take a quick look at the most prominent features of that landscape.

First, you should know that the cord-free world is broadly divided into two main categories:

>> **Over-the-air (OTA) TV:** Live television channels broadcast from a station transmitter. This setup usually requires an antenna, but some streaming services offer live TV channels.

>> **Streaming media:** Television programs — as well as movies, music, podcasts, and other media — made available over the Internet. You can use device apps to access streaming media, but most people use a device such as a smart TV, a set-top box, or a streaming player.

In these two categories, you can have one (or, heck, *all*) of the following viewing experiences:

>> **Watching OTA TV with an antenna:** You want to watch live local stations for free (minus the cost of the antenna, of course). To find out more about this option, see Chapters 3 and 4.

>> **Watching live TV with a streaming device:** You don't want the hassle of setting up a digital antenna. You can find the details in Chapter 5.

>> **Watching live TV with a streaming service:** You're mostly interested in live network broadcasts. To learn more, check out Chapter 5.

>> **Streaming media through a device:** You want to use a device such as a smart TV or a USB stick (such as Amazon Fire TV) to access streaming media through apps. I talk about all this in Chapter 6.

>> **Watching free streaming services:** You're too psychologically scarred from paying scandalously high cable prices and want only free content. I talk about free services in Chapter 8.

>> **Watching paid streaming services:** You want a subscription. Most streaming services require a subscription, so this is your most likely experience. For the details, head to Chapter 9.

Chapter **2**

Snip, Snip. Your 7-Step Plan to Going Cord-Free

This book is your complete guide to cutting the cord of your cable subscription. At this early stage of the journey, your head might be filled with questions: "Can I get there from here?" "Where do I start?" "What's the next step?" "Is it normal to be talking to myself like this?"

Excellent questions all, and this chapter is where you get some answers. Think of this chapter as a map with a big, bright line marking the path to take. Yep, sure, the path meanders a bit, but the seven steps you learn about in this chapter will lead you from wherever you are now to a life that's gloriously cable-free.

As the philosopher once said, the map is not the territory. The steps you read about here are the bird's-eye view of the entire process. You get more details about each step as you progress through the rest of the book.

Step 1: Deciding What You Want to Watch

Psst. Yes, you. Over here. Can I tell you a secret? Can I tell you why some people are happy cord cutters? Sure, everyone's tickled various shades of pink when they first say "so long" to cable. But some folks manage to *stay* happy even after their cable subscription is a distant memory. What's their secret?

You might think the secret is something esoteric involving a complicated setup and a wide variety of streaming services. Nope. In fact, the secret is the opposite of all that. Ready? Here it is:

> Don't try to replicate your current cable lineup.

When people think of cutting the cord, the first path they think of taking is to use streaming services to clone their existing cable channels and content. Is that even possible? Yep, it certainly is, and you're free to go that route if you like. However, duplicating your cable content will almost certainly lead to post-cable dissatisfaction. Why? Because replicating what you have now will probably be more expensive than cable and will definitely be more complicated than cable.

What's the alternative? There are three paths to long-term post-cable happiness:

>> Watch a small number of shows.

>> Subscribe to a small number of streaming services.

>> Go old-school with OTA live TV.

Let's take a closer look at these three options.

Option A: Going with just your "must see" TV shows

Do you have a few shows that you couldn't possibly do without? Or do you have a few content categories — such as movies, live sports, or news — without which cutting the cord is unthinkable?

If you answered "yes" to either question, make a list of these shows or categories. To keep your costs in check, keep the list as short as possible. I'm talking just your A-list shows or content.

When the list is complete, research which streaming service offers each show or content category. These are the services you'll want to subscribe to when the time comes (see Step 5 later in the chapter).

WARNING

The sports category is a tough one. A lot of content is out there, but it tends to be expensive, especially if you have a multisport viewing habit. I talk more about the sports packages offered by streaming services in this book's cheat sheet. (See the Introduction for details.)

Option B: Going with just a few streaming services

Making a list of "must see" shows, as I describe in the preceding section, is the route for folks who have specific TV tastes. If you're more of a generalist, a better method is to choose a small number of streaming services, each of which has both of the following characteristics:

>> **The service offers a ton of content.** Subscribing to even just one or two services that stream lots of TV shows, movies, and other content will make the viewer in you happy because there will always be something entertaining or interesting to watch.

>> **The service doesn't cost a ton of money.** Subscribing to just a few services that don't cost a lot will make the accountant in you happy because your monthly viewing costs will be way less than they were with cable.

For example, you could combine a subscription to Netflix with one or two other general services such as Amazon Prime Video (lots of free content if you're an Amazon Prime member), Apple TV+, or Hulu. That combination will give you more TV shows and movies than you could ever watch, all for a monthly cost that's less than three or four extra-shot, non-fat, soy lattes.

Option C: Going with OTA live TV

Unless you live in an extremely remote area, chances are you can access a decent collection of live over-the-air (OTA) TV channels. This will get you live sports, news, and whatever primetime shows are available on the channels that come your way.

Going with OTA live TV is such an easy and popular post-cable route that I devote three chapters to the topic. To find out everything you need to know, check out Chapters 3, 4, and 5.

TIP

If you think you want to make OTA channels your main source of post-cable entertainment, first check which OTA stations are available in your area. The TV Fool website (www.tvfool.com) can do this for you. I explain how to use it in Chapter 3.

Step 2: Figuring Out the Equipment You Need

No matter how you plan to get your TV jollies after you cut the cord, you're going to need some equipment. What you need depends on what you have and what services you're going to use.

TIP

I should say that you're *probably* going to need some equipment because you can stream many TV shows and movies without buying any or much equipment. That is, most services offer apps that enable you to stream content on your smartphone or tablet. I talk more about this in Chapter 6.

Here's a quick equipment list, with pointers to where I talk in detail about each type later in the book:

>> **Television:** I assume you already have one of these. However, you might be in the market for a new TV to celebrate cutting the cord. Good for you! I talk about what to look for (from a cord-free angle) in Chapter 6.

>> **Internet modem/Wi-Fi router:** If you are currently renting a modem/router from the cable company and will be getting your Internet access from a different provider, you'll

need to buy one of your own. I give you some buying advice in Chapter 7.

>> **HDTV antenna:** You'll need an antenna to bring in OTA signals to your TV. For lots of info about choosing and setting up an HDTV antenna, check out Chapter 4.

>> **Streaming device:** Lots of services, such as Amazon Fire TV and Roku, give you access to tons of streaming services via apps. For example, Figure 2-1 shows some apps available on the main page of Amazon Fire TV. To get those apps, you need a streaming device, such as set-top box, smart TV, or streaming player. Chapter 6 tells you all about these and other streaming knickknacks.

FIGURE 2-1:
Devices such as Amazon Fire TV offer access to content streams via apps.

Step 3: Making a Streaming Services Budget

In Chapter 1, I mention that although cutting the cord to save money is a laudable goal, it's not the only one. Quite a few great reasons to thumb your nose at the cable company exist.

That said, even if saving money is far down your list of reasons to cut the cord, you shouldn't take streaming costs for granted. If you go with only free streaming services (see Chapter 8) or just a few paid services, you probably don't have much to worry about, financially. But note that it's awfully easy and tempting to keep adding new services. It's $10 a month here, $15 a month there, and $5 a month somewhere else. Pretty soon, you're forking out more per month than you did in your cable days.

What's the answer? First, decide how much money you want to spend each month on streaming. That number might be a fraction of your cable costs, or it might be the same as what you now pay for cable. The amount is up to you, but it's important to write that number down and take it seriously to avoid having your streaming costs go through the roof.

What's the best way to respect your streaming cost ceiling? Make a streaming budget. I know, I know: You'd rather get a root canal. I get it. But creating a budget is not that much work. In fact, you already made a good start when you wrote your list of "must-have" content (in Step 1). Add a second column to that list and use it to record the monthly subscription price for each streaming service.

Don't put your pencil down just yet. You also need to write down the extra cash you'll likely pay monthly for Internet service:

>> If your current Internet service is part of your cable bundle, write down how much your Internet access will go up monthly once you ditch the bundle.

>> If your current Internet service comes from another provider, write down how much extra it will cost you per month to increase your monthly data cap to allow for all your extra streaming. (See Chapter 7 for more about the data you'll use while streaming.)

Now what? Two things:

>> Add all your costs and make sure the total comes in under the maximum streaming budget you decided on earlier. Did you go over? It can happen so easily with this stuff. Be ruthless and cut services until you get under your cap. You can so this.

>> Stay on top of your costs. Update your budget as prices change and you add or remove services.

CABLE-LIKE STREAMING SERVICES

Your motivation for ditching cable might not have anything to do with money. It might be that you find your cable company over-lords to be so unbearable that you'd pretty much do anything to get out from under them. I hear you.

If you like the convenience of cable, don't mind paying cable-like prices, and definitely don't want anything to do with the actual cable company, I suggest that you look into subscribing to a *cable-replacement service*. This is a streaming service that offers a large collection of TV channels, movies, and other content, just like the cable company does. It also charges a hefty monthly fee, just like the cable company does. But, crucially, it's not the cable company. If you're interested, you can find more about these services in Chapter 9.

Step 4: Trying Out Lots of Streaming Services — for Free!

If you're not sure about a streaming service, surf to its home page and look around. You'll get a good sense of what the service has to offer. However, even the best streaming service web page is no substitute for using the service itself. But how do you do that without spending a bunch of money, especially for services you might not like?

Two words: *free trial*. Every streaming service worthy of the name offers some kind of free trial (see Figure 2-2). The idea is that you sign up for the service and give them your payment info, but the first payment doesn't go through until after the trial period expires.

This gives you a week, a month, or sometimes even longer to ring the service's bells and blow its whistles. Getting hands-on with the service for a while should be enough to let you know whether you want to continue with it after the trial period expires.

FIGURE 2-2: All major streaming services offer a free trial period.

WARNING

I encourage you to sign up for as many streaming free trials as you have time to check out. However, you should know that some risk is involved because every service takes your payment info when you sign up for the free trial, and you might forget to cancel one or two before the trial period ends. You then end up paying for some services you don't want, which is maddening.

TIP

One easy way to prevent these unwanted and unnecessary charges is to cancel your account immediately after signing up for it. That sounds weird, I know, but most services still give you full and free access until the trial period expires. You can continue with your trial as usual. Then, at the end of the trial, if you like the service, it's no problem asking them to reinstate your subscription.

Step 5: Subscribing to Streaming Services

Back in Step 1, I had you make a short list of shows or content you wanted to take on your post-cable journey. With your list in hand, the next stage on that trip is subscribing to the streaming services that offer those shows or content.

I talk about paid streaming services in just enough detail in Chapter 9, so that's the place to go to get ready to make your streaming commitments. For now, though, I offer the following pointers, which apply to folks who are new to the streaming business:

>> I mention in Step 4 that most streaming services offer a free trial. Even if you're sure you want to become a paid subscriber of a service, it's almost always the right move to sign up the free tryout. In most cases, the trial period is a full month, so you get lots of time to kick the tires to make sure the service is right for you.

>> Most streaming services will give you a bit of a discount if you commit upfront to a longer subscription. A typical deal is to offer 12 months of service for the price of 10 months. Committing to a service for a full year makes financial sense *if* (I repeat, *if*) you're completely sure you'll be happy with the service that entire time. A year's subscription is a bit of a risk because — let's be honest here — you likely don't know what, streaming-wise, will put gas in your tank. Since you're just getting started with all this, it makes sense to forego that discount for now and stick with a monthly subscription. After a few months, if you love what the service has to offer, they'll be happy to take your money in exchange for a longer commitment.

Step 6: Doing a Cord-Cutting Test Drive

Okay, you're almost there. By this leg of the journey, you know which services you want; your equipment is set up and configured; you're sticking to your streaming budget; you've tried (or are trying) a bunch of free trials for streaming services; and you've possibly subscribed to a few services.

Surely this stage of the path is when you get out your metaphorical cord cutters and break free from the cable company, right? Not so fast! Before you launch your post-cable life, give everything a test drive to make sure it all works as advertised. Call this your pre-post-cable life.

For this test drive to work, you first need to do two things:

>> Connect and configure all your streaming equipment.

>> Disconnect the cable TV cord. To be clear, I'm talking here about the actual cord that attaches to your TV.

Why disconnect the cord? I suppose it's not strictly necessary. But if you want this test drive to accurately reflect what your TV-watching life will be like AC (after cable), cable TV itself must be off-limits for the duration of the test. If you leave the cable connected to your TV then — admit it — you won't be able to resist the temptation to switch inputs and check out something available on cable that won't be available when you go cable-free for real.

Now you're ready to test-drive your streaming life. For the next week or two (or however long you think you need to give the new setup a fair shake), try out your services, configure settings, and run experiments. This is the time to really push your services to the limit to make sure everything works.

Note that this experiment might not be a perfect test of what you'll experience on the other side of cable. In particular, after you cut the cord, you might decide to opt for a faster Internet connection speed or to upgrade your Wi-Fi. If you plan on doing either (or both), bear in mind that your test-drive experience will be a little slower than your experience after you've made your upgrades.

Step 7: Cutting the Cord — Woo-Hoo!

Well, I'm sure you've been dreaming about this day for a long time, and it's finally here: It's time to call your cable company and cancel your service. It's time, in short, to cut that cord!

I wish I could tell you that the ensuing phone call will be simple, take just a few minutes of your time, and be completely stress-free. Nope, sorry, but this is the cable company I'm talking about. *Nothing* they do is ever simple, quick, or relaxing.

That's fine. You've got this. To help, the rest of this chapter takes you through the process and offers plenty of tips and advice for getting out the other end with your sanity (relatively) intact.

Check your contract

Dig out your cable subscription contract, if you still have it. If you don't, you should be able to find it online using the cable company's website. Peruse the contract for a clause that states that the company can charge a fee for an account cancellation. You need to know this going in so that it doesn't come as a shock if the rep informs you, which might cause you to lose your cool — or even change your mind about cancelling.

The good news here is that you might have some leverage for asking the company to reduce — or, ideally, waive — the fee if you're going to stick with them for Internet access.

Check your bandwidth history

I assume that your cable company or Internet service provider offers an online customer portal that enables you to access your account data and history. If so, go ahead and sign in, and then navigate to the page that shows your Internet bandwidth history. This page should tell you not only how much bandwidth you used each month but also how that usage compared to the usage cap on your account.

Why do you need this info? If the data shows that you're consistently close to or even over your bandwidth ceiling most months, you're certainly going to blow that ceiling out of the water every month when you start streaming full-time. Knowing this will help you negotiate a new (read: much higher) bandwidth ceiling with whatever company you use for Internet access going forward.

Gather everything you need

I know, I know, it's just a phone call. What could you possibly need to gather? You'd be surprised:

>> Have your cable company account number handy.

>> Better yet, have a copy of a recent cable bill nearby.

>> Have some paper and a pen or pencil at hand so you can jot down a few notes as needed during the call.

>> Pour yourself some water, a coffee, or your beverage of choice.

>> You might end up on hold for long stretches, so bring along something to do: a crossword, the novel you're reading, a cable rep voodoo doll, or whatever.

Get ready mentally

Before you even pick up your phone, remember that one of the reasons many people want to sever their relationship with their cable provider is the atrocious customer service I mention in Chapter 1. When the cable company treats even its *paying* customers with disrespect and disdain, how do you think they're going to treat you, the account canceler?

So, before dialing, give yourself a talking-to. Remind yourself that the ensuing conversation will almost certainly be frustrating and stressful. Remind yourself that you're making a solid decision here. You did your research, made your budget, ran your trials and tests. You're ready. Remind yourself that, ultimately, this is *your* decision, so it doesn't matter what some customer service rep has to say about it.

Know the process

Every company has a different procedure for handling customers who want to cancel their account, so it's hard to be definitive on the exact steps of the phone call. However, the progress of the call will probably go something like this:

1. The rep answers the call.

TIP

   The customer service rep will almost always answer the call by stating his or her name. Be sure to write down that name so you can refer to it later.

   You state your business up front:

"Hello. I want to cancel my cable TV subscription effective immediately."

2. The rep will probably ask you why you want to cancel.

This is the first stress point in the call because, secretly, you were hoping the rep would say something like "Why, of course, I'll do that for you right away." Instead, you're forced to justify your decision. That's fine, just keep your answer short and to the point:

"I've decided I no longer need cable TV and therefore I no longer need your company cable TV services."

3. The rep will now likely ask if there is anything the company can do to improve your service.

This is another stress inducer because, again, you're not getting what you want right away. Shrug it off and restate your intention:

"No, thank you. I just want to cancel my cable TV account."

4. The rep will now either offer you a special promotion or some other deal if you stay with the company. Alternatively, the rep will pass you over to the dreaded customer retention agent, a cable company employee whose job it is to convince people like you to not cancel their accounts.

Whatever anyone offers or says, remain calm, be polite, and stay firm:

"I appreciate the offer, but no thanks. I'd like to cancel my cable TV account, please."

TIP

If you still want to use the cable company for your Internet access, at some point during the call you should ask the rep about that. Find out what the monthly cost is for an Internet-only account, particularly one that has a suitable bandwidth cap (or even unlimited bandwidth). Feel free to ask the rep if the company can offer you a deal on Internet access.

5. The rep — particularly the retention agent, who probably gets a bonus for each retained customer — might double down and either make you a better offer or lecture you on the alleged evils of cord cutting and streaming services.

Let the person have his or her say. What do you care? When the lecture is over, politely but firmly restate your intention:

"Thank you. I'd like you to cancel my cable TV account now, please."

6. If the rep is particularly nasty, you'll get put on hold — for a very long time. Why? Because the rep is hoping that you'll lose patience and hang up!

Remind yourself that this sort of treatment exemplifies why you want the cable company out of your life. Stay patient, do your crossword, read your book, or find some other pleasant way to pass the time, and wait the person out.

7. At long last, the agent will cancel your account.

Say "Yes!" to yourself, pump your fist a few times, and then get back to the business at hand. Before you hang up, ask the rep the following questions:

- What is the confirmation number, indicating that my service is cancelled?
- When does the cancellation takes effect?
- Do I have to pay any outstanding account balances or fees?
- What equipment, if any, do I have to return (see the next section), how long do I have to return it, and what address do I use?

8. Hang up and revel in your newfound freedom for as long as you want. You've earned it.

Return all equipment

You probably have some cable company property in your house in the form of an Internet modem/router or a cable TV set-top box or both. As soon as possible after you cancel your account, return this equipment to the cable company. From the cable company agent who cancelled your account, you should know how long you have to get the equipment back to them. However, don't wait as long as that because you don't want to give the company any excuse to hit you with charges related to missing or late equipment returns. The sooner you get that equipment back — be sure to take the equipment back in person and to get a receipt for it — the sooner you can say the cord is now officially, blissfully, cut.

2

Cord Cutting Made Easy with Over-the-Air TV

IN THIS CHAPTER

» **Understanding this over-the-air TV stuff**

» **Debating the pros and cons of over-the-air TV**

» **Learning the ABCs of OTA**

» **Finding out which channels you can access**

» **Getting your over-the-air need-to-know**

Chapter **3**

Understanding Over-the-Air TV

I f you have just cut the cord — or are planning to soon — and are wondering where to go next, how about something that's almost as free and easy as breathing the air? I thought that might catch your attention. What could possibly be so cheap and so simple that it's comparable to inhalation? I speak of a television signal called over-the-air, which is the subject of this chapter. As I show in the pages that follow, over-the-air TV is freely available assuming you live in or are not terribly remote from a major urban center. With a modest investment of equipment, those signals are yours for the viewing, no questions or fees asked.

In this chapter, you discover what over-the-air TV is all about. You learn why it's the first — and for many, the only — stop in their post-cable travels. You explore how over-the-air TV works and investigate some exciting new developments in the over-the-air world. And perhaps most importantly, you learn how to find nearby over-the-air channels, which will help you decide if over-the-air is worth checking out. Will you be over-the-moon about over-the-air? Let's find out.

OTA? OTT? Live TV? What on Earth Is Everyone Talking About?

When you're watching TV, you're tuning it to one of the following:

>> **Live:** Content that is happening now (such as a live sports event) or being broadcast to everyone at the same time.

>> **On-demand:** Content that has been prerecorded. You decide when you want to watch it.

When you watch something live on TV, you're watching a specific live television signal. That explanation sounds straightforward enough, but there are actually four different types of live television signal:

Signal type	How the signal gets to your home
Cable TV	A cable
Satellite TV	An orbiting satellite
Broadcast TV	Radio waves from a TV station transmitter
Internet TV	A video stream over the Internet

I assume that you have cut the cord (or will do so soon) and no longer have a cable TV subscription, and that you don't use satellite TV as a substitute. That leaves broadcast and Internet as live TV alternatives, which also go by the following abbreviations:

>> **OTA (over-the-air):** Another name for broadcast TV. The signal is literally sent through the air.

>> **OTT (over-the-top):** Another name for Internet TV. The signal bypasses (so, in a sense, jumps over the top of) cable, satellite, and broadcast receivers.

In this chapter and in Chapter 4, I talk about over-the-air TV. In Chapter 5, I discuss watching both over-the-air TV and live TV streamed over the Internet.

What is over-the-air TV?

Over-the-air TV — which also goes by the monikers *broadcast TV*, *linear TV*, and *terrestrial TV* — refers to television signals sent into the air via a TV station's transmission tower. As I discuss a bit later, those transmission towers are Earth-bound (as opposed to, say, an orbiting satellite), hence the name *terrestrial* TV.

To view the signal from nearby transmission towers, you need an antenna, which I discuss in detail in Chapter 4. You then connect the antenna to your TV and you're good to go.

TECHNICAL STUFF

LINEAR VERSUS NON-LINEAR TV

Why is over-the-air TV called *linear TV*? Because when you watch over-the-air TV, you tune in to a particular program at the time it's scheduled to air, watch it to the end (presumably), watch the next show at its scheduled time, and so on. The shows you watch are, in a sense, lined up one after the other, so your viewing experience is a *linear* one.

By contrast, shows available via over-the-top TV are always available, so you can watch them whenever the mood strikes. You can watch half an episode of a show, switch to a different show, return to the first show, and then binge-watch a third show until you fall asleep on the couch. You can jump around willy-nilly, so your viewing experience is a *non-linear* one.

Good for what, you ask? The answer depends on where you live and how close you are to an urban center. Generally speaking, you can expect some or all of the following when you tune in to over-the-air TV:

>> Newscasts that feature coverage of local news, sports, and weather: Here, *local* refers to whatever urban center is host to the station transmitting the over-the-air signal.

>> Live local sporting events.

>> The full lineup of shows from most major broadcasters. In the United States, you should receive most or all of the following stations: ABC, CBS, the CW, Fox, NBC, and PBS. Other major broadcasters include CBC and CTV in Canada and the BBC in Great Britain.

>> One or more national over-the-air broadcast networks, such as Home Shopping Network, ION Television, MeTV, or Univision.

TIP

For a complete list of the over-the-air TV networks available in the United States, check out the following Wikipedia article: https://en.wikipedia.org/wiki/List_of_United_States_over-the-air_television_networks.

>> Community-focused content such as public access channels, locally produced shows (live or prerecorded), and locally broadcast stations.

REMEMBER

The number and variety of available channels depends on where you live. As a general yardstick, locations in or near major cities can often get up to 50 channels; locations in or near mid-sized cities might see around 35 channels; and locations in or near small cities might get around 15 channels. These numbers assume ideal conditions, so in the real world most folks get fewer channels.

The pros and cons of over-the-air TV

The pros of OTA TV aren't hard to find: free content, access to local news and sports coverage, and a relatively inexpensive and simple hardware setup.

REMEMBER

Here's another item to add to the "pro" side of the OTA TV ledger: excellent picture quality, particularly if you're used to cable TV. The signals that come via cable are massively compressed so that the cable company can transmit more channels at once. That compression greatly reduces signal quality. OTA TV signals, by contrast, are only compressed a little, so the picture looks *much* better than it does on cable.

Are there any negatives to consider? Yep, a few:

>> **One word: commercials:** Broadcast TV is free because it's supported by commercials, *lots* of commercials. Be prepared to make steady and heavy use of your TV remote's Mute button.

>> **Lack of viewing freedom:** Unless you pay extra for a digital video recorder (DVR) that works with OTA TV (see Chapter 5), you have no choice but to watch OTA shows only when they air.

>> **Limited channels:** Depending on where you live, your OTA equipment might pick up a few dozen broadcasting stations or just a few. Either way, you're facing a limited TV lineup compared to what you had in your cable days.

>> **No cable channels:** By definition, OTA TV doesn't include any channels that broadcast via cable TV, so that means no HBO, no Showtime, no ESPN, and no CNN.

>> **No technical support:** With an OTA TV setup, you're on your own when it comes to troubleshooting problems. Of course, if you're just coming from cable, you're used to having non-existent technical support!

>> **No channel guide:** In a basic OTA setup, usually you can find out what's on only by flipping through the channels. I know: *so* primitive! Fortunately, you can peruse online TV listings, and for a bit of extra cash each month, you can add a channel guide to your configuration. Note, too, that an OTA DVR includes a channel guide, as do some TVs.

TECHNICAL STUFF

WHAT IS NEXT GEN TV?

If you live in the United States, in the not-too-distant future you'll have the opportunity to upgrade your over-the-air TV experience to something called *Next Gen TV*. This feature also goes by the far less catchy name *ATSC 3.0*, where ATSC is the Advanced Television Systems Committee, which makes broadcast TV standards for North America. Next Gen TV is a relatively new broadcast technology that improves on the existing ATSC 1.0 standard. (In case you're wondering: Yep, there was an ATSC 2.0, but it was a for-nerds-only release and can be safely ignored by the likes of you and me.)

Next Gen TV offers much higher picture quality, improved signal reception (especially for indoor antennas), and a vague promise of interactive features (which, if the past is any indication, will be lame and of interest to only marketers).

More problematically, Next Gen TV also promises targeted advertising, meaning commercials specific to where you live and your viewing preferences. The ATSC promotes targeted advertising as an improvement, but in my experience such advertising either isn't very targeted or works all too well and becomes creepy (like the ads that follow you around the web). The implication here, too, is that broadcasters will be able to collect data on your viewing habits, which has major privacy implications. However you look at it, the targeted advertising feature of ATSC 3.0 isn't much of an improvement.

What should you do about Next Gen TV now? Probably not much. Few stations broadcast Next Gen TV signals, although more stations are going live all the time. The biggest hurdle currently is hardware. Any antenna can pick up Next Gen TV signals, but you need an ATSC 3.0 tuner to view those signals on a TV. You can shop around for a TV that has a built-in ATSC 3.0 tuner or for a set-top box that can convert an ATSC 3.0 signal to something your TV can display. Be warned, however, that prices for ATSC 3.0-compatible hardware are expensive.

If you want to keep up with the latest ATSC 3.0 deployments in the United States (and South Korea, the only non-North American market that uses ATSC standards), keep an eye on the following page: www.atsc.org/nextgen-tv/deployments/.

How Over-the-Air TV Works

If you ask a fish, "How's the water?" the fish is likely to reply, "What's water?" because, from the fish's point of view, water is everywhere, all the time. To the fish, water just *is*. Over-the-air TV signals have the same everywhere, all-the-time characteristics, so you might be forgiven for thinking, fish-like, that over-the-air TV just *is*. Of course, that's not true, but from where do all those signals originate? Let's take a look.

The television station

From the perspective of over-the-air TV, all signals begin at a television station. There are four main kinds of stations:

Station type	What it is
Owned and operated (O&O)	Is the property of, and is run by, a national broadcast network (such as one of the so-called Big Five in the United States: ABC, CBS, the CW, Fox, or NBC). O&O stations tend to be located only in major cities.
Network affiliate	Carries some or all of the programs broadcast by a particular national broadcast network, but the station is independently owned and operated.
Member station	Is part of a collection of stations that together own the network. The main example in the US is the Public Broadcasting Service (PBS). Each member station is independently owned and operated.
Independent	Is not affiliated with any national broadcast network and is independently owned and operated.

Whatever the station type, the TV content supplied by the station will be comprised of either or both of the following:

>> **Internal content generated in the station itself:** This content could be a live local newscast or similar live-to-air programming or prerecorded shows, such as reruns of previously aired local shows.

>> **External content transmitted to the station:** This content could be shows supplied by a parent national broadcast network — the so-called *network feed* — or live coverage of local happenings, such as sports or entertainment events.

The transmission tower

The station converts the content to a digital signal. Most digital signals nowadays are in high-definition (HD) format, although some stations might broadcast using the standard-definition (SD) format. In the future, many digital television signals will be ultra-high definition (UHD). (UHD is part of the ATSC 3.0 stuff I noted earlier in the "What Is Next Gen TV?" sidebar). Is all this gobbledygook to you? Then see the upcoming sidebar "SD? HD? UHD? What Is This Stuff?" for the 411.

That signal includes both the video and synchronized audio, as well as data such as the channel number, associated network (if any), and closed captioning information. The station passes the digital signal along to a television transmitter. The transmitter is usually a tall tower installed in a high position, such as on the roof of a building or the top of a hill. The broadcast tower then transmits the digital television signal as radio waves in all directions. This is the "over-the-air" part of over-the-air TV.

For a *full-power* station, the transmission tower will have a range of between 50 and 80 miles (80 to 128 kilometers); for a *low-power* station, the transmission tower's range will be between 15 and 30 miles (24 to 48 kilometers).

The antenna and tuner

Okay, so myriad television signals are flying around your head as you read these words. How do you redirect those signals from the air out there to your TV in here? To view over-the-air signals in your home, you need three things:

>> An indoor or outdoor antenna to pick up the signals. See Chapter 4 to learn everything you need to know about selecting and installing an antenna to pick up over-the-air signals.

TECHNICAL STUFF

SD? HD? UHD? WHAT IS THIS STUFF?

Formats such as HD and SD are shorthand ways to refers to the *resolution* of a digital signal. The resolution determines, in a sense, how sharp the video portion of the signal will appear on your TV. To understand how this works, first know that your TV screen is comprised of a grid of tiny points of light. Each point is called a *pixel* (a smooshing together of the phrase *picture element*). Every pixel displays a constantly changing light, the color of which is some amalgam of red, green, and blue.

Your TV's pixel grid arranges the pixels in rows and columns. The number of pixels in each row is called the *horizontal resolution*; the number of pixels in each column is called the *vertical resolution*. In a nutshell, the bigger the numbers for the horizontal and vertical resolution, the sharper the picture.

Standard-definition (SD) TVs have a horizontal resolution of 720 pixels and a vertical resolution of either 486 pixels or 576 pixels. These resolutions are usually written as 720 x 486 and 720 x 576, respectively. These resolution values mean that either 349,920 or 424,720 total pixels, respectively, are available for each video frame.

That sounds like a lot, but the high-definition (HD or HDTV) resolution is 1920 x 1080, which means 2,073,600 pixels. That's five or six times the SD value, which is why HD looks so much better than SD. A second HD format is 1280 x 720.

The ultra-high–definition (UHD or UHDTV) resolution is 3840 x 2160, so TVs that support this format are teeming with 8,294,400 pixels, which is four times what you get with HD. The horizontal resolution of 3,840 pixels is why lots of people refer to this format as 4K UHD.

Just so you know, TVs exist that support resolutions of 7630 x 4320 (32,961,600 pixels total), a format known as UHD-2 or 8K UHD.

>> A tuner integrated into a television or a similar device (such as a DVR). Note that the tuner must be capable of interpreting the television signal picked up by the antenna. For example, most television signals are transmitted in HD format, so your TV (or whatever) must have an HDTV tuner.

>> A coaxial cable that brings the signal from the antenna to the tuner.

Figure 3-1 shows how the whole over-the-air TV process works.

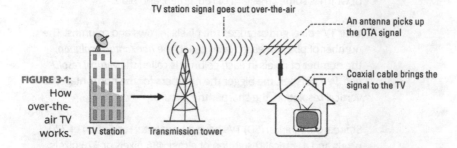

TV station signal goes out over-the-air

An antenna picks up the OTA signal

Coaxial cable brings the signal to the TV

FIGURE 3-1: How over-the-air TV works.

TV station   Transmission tower

What Channels Can You Access?

Over-the-air TV itself is free, but as I mention in the preceding section, you need a few pieces of equipment to bring the signal into your home, and those bits of hardware aren't free. You probably have a TV and coax cable, so your major purchase here is the antenna. I give a few antenna buying tips in Chapter 4.

Before you get to that, however, you might want to know if going any further with this over-the-air TV stuff is even worthwhile. In other words, how many stations can you get and what are they? The number of channels you can access depends on the strength of the over-the-air signals in your neck of the woods. To find out what stations are available, you can use several tools.

Understanding the factors that affect signal strength

As mentioned, TV station transmitters have ranges as low as 15 miles (24 kilometers) for low-power stations and as high as 80 miles (128 kilometers) for full-power stations. However, the power of the original signal is only one factor that determines how strong the signal is by the time it gets to you. Here are some other factors that come into play:

>> **Your location:** Most TV station transmitters are situated in or near urban centers. The bigger the city, the more transmitters it will host. If you live in a city, you'll likely receive a strong signal from most of that city's TV transmitters. The farther away you live from the city, the fewer signals you'll pick up and the weaker those signals will be.

>> **Your antenna's range:** If you live 50 miles (80 kilometers) from a transmission tower and that tower is connected to a low-power station with a maximum range of 20 miles (32 kilometers), you're out of luck, right? Yes, if your antenna's range is just a few miles, as it is with most cheap indoor antennas. However, costlier outdoor antennas often have ranges up to 70 miles (112 kilometers), and such an antenna would pick up the low-power station no problem.

>> **Your antenna's direction:** Cheaper OTA antennas are *unidirectional*, which means they pick up signals from only a single direction. If you have such an antenna, you'll get a greater signal strength from a particular transmission tower if your antenna is pointed at that tower.

>> **The line-of-sight between your antenna and local transmission towers:** For the best signal quality, your antenna requires a line-of-sight with the broadcast tower. *Line-of-sight* means you could draw a straight line between your antenna and a transmission tower without going through a tall object such as a building or a hill. Line-of-sight might be a problem is you live in a valley or if your location is surrounded by tall trees.

>> **Objects that interfere with the broadcast signal:** The signals you receive can be degraded by any number of modern technologies, especially the presence of nearby cellular network (especially LTE) towers or power lines.

REMEMBER

Having no line-of-sight with a broadcast tower or having nearby objects that interfere with the signal can produce weak or poor reception. However, more often than not, you simply don't receive the station at all, a phenomenon known in the trade as the *digital cliff*.

TIP

If you have a nearby cellular network tower that might cause interference, you can install an *LTE filter* between your antenna and your TV. This device filters out interference from LTE cellular signals to give you better over-the-air TV broadcast reception.

>> **Today's weather:** This one sounds like a joke, but it's true. Extreme weather such as heavy rain or snow, a peasouper of a fog, and wild swings in temperature can wreak havoc on a broadcast signal.

Now that you know what factors can influence a signal, you're ready to start looking around to see what's available in your area.

TIP

If you live in Great Britain, you can use an antenna — sorry, I mean an *aerial* — to access free TV content supplied by the Freeview platform. Freeview supports more than 80 channels, including the BBC, Sky, ITV, and Channel 4. To see which channels are available in your area, go to www.freeview.co.uk/check-channels-home.

Checking the FCC's digital TV reception maps

The Federal Communications Commission (FCC) regulates television (as well as radio, satellite, and cable) for the United States. The FCC website offers a DTV Reception Maps page where you can use your address or current location to bring up a list of nearby digital TV stations.

REMEMBER

As you might imagine, since the FCC is a department of the US government, the DTV Reception Maps tool only works for US-based locations.

Here's how the DTV Reception Maps tool works:

1. **Use your favorite web browser to surf to** www.fcc.gov/media/engineering/dtvmaps.

2. **Use the text box to enter your address.**

For the best results, include as much information as possible: street number and name, city, state, and ZIP code.

Alternatively, you can submit your current location by clicking the Go to My Location! button. Note that your web browser might ask if it's okay to supply the page with your current location. If you're cool with that, be sure to click Allow or OK or whatever to give your permission. If you go this route, skip Step 3.

3. **Click the Go! button.**

The website displays a table named DTV Coverage that shows the available over-the-air TV stations in your area, as shown in Figure 3-2. This data gives two indicators of signal strength — color and signal bars (in the first column). Here's how to interpret these indicators:

- *Strong signal:* The station info has a green background and four bars.

- *Moderate signal:* The station has a yellow background and three bars.

- *Weak signal:* The station info has a brown background and two bars.

- *No signal:* The station info has a red background and you see an X instead of bars.

WARNING

The DTV Coverage signal strength results are based on the assumption that you're using an outdoor antenna mounted 30 feet above the ground. If you're using an indoor antenna or an outdoor antenna mounted lower than 30 feet, your signals will be weaker than what the FCC shows.

You also see your location marked with a pin on a map (not shown in Figure 3-2).

Callsign	Network	Chn	Band	IA
Click on callsign for detail				
WJLP	MeTV	33	Lo-V	
WABC-TV	ABC	7	Hi-V	
WCBS-TV	CBS	2	UHF	R
WNJU	TELE	47	UHF	R
WNET	PBS	13	Hi-V	R
WPIX	CW	11	Hi-V	
WNYE-TV	ETV	25	UHF	
WFUT-DT	UNIM	68	UHF	R
WPXN-TV	ION	31	UHF	R
WDVB-CD			UHF	R
WLIW	PBS	21	UHF	R
WNYW	FOX	5	UHF	R
WWOR-TV	MY N	9	UHF	R
WMBC-TV	IND	63	UHF	
WNJB	PBS	58	Hi-V	
WLNY-TV	IND	55	UHF	R
WPVI-TV	ABC	6	Lo-V	
WDPN-TV	NBC	2	Lo-V	

FIGURE 3-2: A list of nearby over-the-air TV stations and their relative signal strengths.

4. **To try a different location on the map, click and drag the pin to the location you want.**

   The page updates the DTV Coverage table for the new location.

Using TV Fool's TV signal locator

Probably the best tool around for checking out the available over-the-air TV signals in your neighborhood is the TV signal locator tool offered by the TV Fool website (www.tvfool.com). It works in both the US and Canada and gives you tons of information about each station signal.

To use the TV Signal Locator tool, follow these steps:

1. **Point your web browser to www.tvfool.com.**

2. **In the Tools section on the left, click TV Signal Locator.**

   The TV Signal Locator page appears.

3. **Select the Address radio button and enter your location info using the Address, City, State/Province, and Zip/Postal code text boxes.**

   If you're not comfortable giving TV Fool your full address, you can use a more anonymous method that uses latitude and longitude. Simply select the Coordinates radio button and specify your location using the Latitude and Longitude text boxes.

TIP

Okay, I hear you ask, how do I get these coordinates? The easiest way is to surf to Google Maps (https://maps.google.com), and run a search for the location you want to use. Right-click the pin that shows your location. The top of the shortcut menu shows, in order, your latitude and longitude.

4. **(Optional) Use the Antenna Height text box to enter the height, in feet, that your antenna sits above ground level.**

If you don't have your antenna yet and aren't sure where you want to mount it, leave this box blank. In that case, TV Fool uses a default value of 10 feet.

5. **Click Find Local Channels.**

TV Fool mulls things over for a few seconds, and then returns the TV signal analysis report, as shown in Figure 3-3.

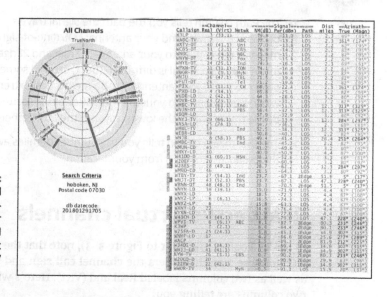

FIGURE 3-3: A typical over-the-air channel report from TV Fool.

The main part of the report is divided into two sections:

>> The radar chart on the left tells you the direction of each signal.

>> The table on the right lists the available signals for your location.

The table lists the stations in descending order of signal strength. The results use the following color code:

Color	Signal strength
Green	The signal is strong enough to be picked up by a basic indoor antenna.
Yellow	The signal is less strong, so it probably requires a larger antenna, a higher antenna (such as an attic-mounted antenna), or both.
Red	The signal is relatively weak, so you most likely require a roof-mounted outdoor antenna to pick up these signals.
Gray	The signal is too weak to pick up with even with a large, outdoor antenna.

TIP

The table has a Signal section that provides two useful values:

>> The Path column tells you the path the signal travels between the tower and your antenna: LOS (line-of-sight) means a direct path to your antenna; 1Edge and 2Edge mean the radio signal is diffracted (bent) around some object to get to your antenna; Tropo means the (extremely weak) signal comes to your antenna after being scattered by the troposphere (the lowest layer of the atmosphere).

>> The Dist miles column tells you the number of miles away the broadcast tower is from your location.

Real versus virtual channels

In the TV Fool report (refer to Figure 3-3), note that the table has a Channel section that offers the channel call sign and network, as well as two columns labeled Real and (Virt). Here's what these two columns are telling you:

>> **Real:** The channel number corresponding to the frequency at which the channel is broadcast — the *transmit channel* or *RF channel* (see Chapter 4 for more about this).

>> **(Virt):** The *virtual channel* number, which is the number you tune into using your TV remote.

Why are they different? Back when all channels were analog, many stations used their channel number as part of their brand. For example, Figure 3-4 shows the logo for WNBC in New York, which in the days of analog TV aired on channel 4. As you can see, the number 4 plays a prominent role in the logo.

FIGURE 3-4:
Many TV
stations
use their
channel
number
in their
branding.

The problem is, when TV signals switched to digital, the digital channel number assigned to some stations wasn't the same as the station's analog channel number. WNBC, for example, was assigned the digital channel 28. Of course, the marketing folks at WNBC and hundreds of other stations didn't want to throw out years of branding. The workaround was to enable these stations to create virtual channels corresponding to their original analog channel numbers. WNBC, for example, uses virtual channel 4.1. When folks in New York tune to channel 4.1, they see WNBC, even though the station is being broadcast on a completely different "real" channel.

Mapping digital channels to branded channel numbers is only part of the virtual channel story. One of the advantages that broadcasters gained with the switch to digital TV was the capability of transmitting multiple channels on a single broadcast transmission, which is why virtual channels use decimals, such as the 4.1 for WNBC in New York. The .1 means this is the first signal transmitted on this frequency. As it happens, the transmitter for WNBC also includes two other channels: Cozi TV, which airs on virtual channel 4.2, and NBC LX, which airs on virtual channel 4.3.

TIP

One of the problems with the TV Fool output is that it doesn't tell you about any of the extra channels provided by a given broadcast transmission. To see these extra channels, try an alternative tool such as AntennaWeb (https://antennaweb.org/).

Chapter **4**

Choosing and Setting Up OTA Hardware

I f you want to watch live local broadcasts, you don't necessarily have to invest in new equipment. As I explain in Chapter 5, streaming services enable you to watch live TV using your existing Internet connection and Wi-Fi network. These services are convenient but not cheap.

What if you want to watch live local broadcasts for *free?* Then you need to go the over-the-air route, and that route requires some new electronics. Fortunately, you're not looking at a substantial investment here. Sure, you could spend big bucks if you want to, but a modest investment is all that's required for most people. In this chapter, you learn what equipment is required and how to make smart purchases.

What about getting the equipment installed? With some exceptions. The good news is that you don't need to bring in an expert to hook up everything. Most of the connections are straightforward, and I tell you everything you need to know in this chapter.

What Equipment Do You Need?

Before you get started, be sure to check that you can even receive OTA TV signals in your area. This won't be an issue if you live in or near a major urban center, but the farther away from civilization you reside, the fewer over-the-air TV signals you'll receive. There's no point in investigating OTA equipment if no OTA signals are in the air where you live. See Chapter 3 to learn how to check what channels are available in your locale.

That said, let's take a quick look at what's required to receive and watch over-the-air TV broadcasts. I start with the watching part, which you can do using any of the following devices:

>> **TV:** Most people watch over-the-air programs on a TV. You don't need anything fancy, just a TV with an F connector (also called an RF port) for the antenna, as shown in Figure 4-1, and a built-in digital tuner. Almost every TV sold since early 2007 has such a tuner. If you have an older TV without a digital tuner, you'll need to spring for an external digital TV tuner or DVR.

WARNING

If you're looking to purchase a new TV, be careful. Some manufacturers are selling devices that look a lot like TVs but are marketed as displays. What's the diff? A display has no F connector for your antenna's coaxial cable, nor does it have a tuner! Check the specs before buying to make sure the set has both an F connector and a built-in digital tuner.

>> **Computer:** Computer monitors don't come with digital TV tuners, so you need an external digital TV tuner box that can connect to your computer monitor (say, using an HDMI cable).

FIGURE 4-1:
Your HDTV
antenna
connects
to your
TV's F
connector.

>> **Mobile device:** You need a digital TV tuner. If your mobile device runs Android, you can buy a tuner that connects directly to your device via USB. If you're using an iOS device, look for a digital tuner that can connect to and broadcast over your Wi-Fi network (see, for example, the HDHomeRun family of tuners from SiliconDust).

To receive over-the-air TV signals, you need some or (rarely) all of the following:

>> **HDTV antenna:** Corrals the over-the-air signals whooshing by. An antenna is the only bit of equipment you definitely need. See the next section, "Choosing an OTA Digital Antenna," to learn some buying tips and suggestions.

>> **Signal amplifier:** Boosts the antenna signal. If you live far from an urban center and the resulting over-the-air TV signals are weak, a signal amplifier can help improve the reception into something watchable.

WARNING

If you live in an area that gets plenty of strong signals, you might be tempted to add a signal amplifier anyway as a way of boosting weaker signals. Paradoxically, adding the amplifier will probably make *all* your signals worse because your already strong signals will become *too* strong and the tuner won't be able to handle these overly boosted signals — a phenomenon known as *overdriving* the tuner.

TECHNICAL STUFF

UHF VERSUS VHF: LET'S TALK RADIO

Over-the-air TV broadcast transmissions are radio signals. In particular, OTA transmissions use two different radio frequency bands:

- **Very high frequency (VHF):** VHF radio frequencies range from 30 to 300 megahertz (MHz). The specific ranges used by TV signals vary depending on the country. In the United States and Canada, the VHF range for TV transmissions is between 54 and 82 MHz for channels 2 through 6 (sometimes called the *low-VHF* or *lo-V* region) and between 174 MHz and 210 MHz for channels 7 through 13 (sometimes called the *high-VHF* or *hi-V* region).

- **Ultra high frequency (UHF):** UHF radio frequencies range from 300 MHz and 3 gigahertz (GHz). As with VHF, the specific range used by TV signals varies by country. In the United States and Canada, the UHF range for TV transmissions is between 470 MHz and 692 MHz for channels 14 through 51 (with channel 37 reserved for radio astronomy).

The VHF and UHF frequency bands are home to channels 2 through 51. These channels are the *transmit channels* or *RF channels*, meaning they correspond to physical radio signal transmissions. Remember, however, that most RF channels have a corresponding virtual channel number (see Chapter 3), and it's the virtual channel number you tune to on your TV.

>> **Signal attenuator:** Reduces the antenna signal. Why on Earth would you ever want to do that? If you have a broadcast tower in your neighborhood — within, say, about 5 miles (8 kilometers) — the signal might be so strong that you overdrive your tuner. Connecting an attenuator can fix that problem.

>> **LTE filter:** Removes the interference caused by nearby LTE cellular network towers and devices. This doohickey is optional, but if you live in an urban center with all its attendant cellular noise, the $10 or $15 investment is probably worthwhile.

Some HDTV antennas come with a built-in LTE filter, so you might want to add that feature to your antenna wish list.

» **Coaxial cable:** Connects the antenna directly to your TV or to an external digital TV tuner (or DVR) if you're using an older TV or a computer or mobile device. (See the section "Connecting Your Antenna" for more on getting everything hooked up.)

» **Coaxial cable splitter:** Splits the signal from your antenna so you can route the signal to multiple devices. Most splitters have two, three, or four outputs. Note that you lose signal strength when you use a splitter. The better splitters sacrifice only 3.5db (decibels) of power (compared to 7db for some splitters), so check the specs.

If you get a splitter and find that you're not using one or more of the outputs, be sure to plug every unused output with a little device called a *terminator* (insert Arnold Schwarzenegger joke here) to prevent signals from leaking from the open outputs.

Here are a few tips for buying coax cable:

» *Don't buy a cable if you don't have to.* Many antennas come with their own coaxial cable, so choose your antenna first, and then shop for a coaxial cable if the antenna doesn't provide one. That said, there's a good chance the coaxial cable that comes with the antenna will be either too short or too cheap (see the next item in this list). In that case, make sure the cable is detachable from the antenna, which will enable you to substitute your own (better quality) cable.

» *Don't be cheap.* Inexpensive cables are tempting, pricewise, but they're usually poorly shielded, which makes them susceptible to signal loss through leakage and to interference from nearby devices. Both signal leakage and external interference can degrade the OTA signals running through the cable. Look for coaxial cables that have three layers of shielding (*triple shield*) or, even better, four layers of shielding (*quad shield*). Also, be sure to buy RG-6 (or RG6) coaxial cable, which has less signal loss overall than the older RG-59 (RG59) cables.

>> *Get enough (but not too much).* If you already have at least an idea where you'll install your antenna, take a rough measurement of the distance from that position to your TV. (For accuracy, be sure to measure down the wall and across the floor, not in a straight line from antenna to TV.) Add a few feet for good measure and start shopping. Why not get a super-long cable, just to be safe? Because over-the-air TV signals degrade as they travel along the cable, so the longer the cable, the more degraded the signal when it finally emerges at your TV.

>> *Get cable that can stand up to the weather.* If you're going to install your antenna outdoors, be sure to get coaxial cables designed for outdoor use. Since you'll almost certainly have at least one cable-to-cable connection outside, look for cables with a rubber boot at each end to weatherproof the connection.

Choosing an OTA Digital Antenna

Okay, enough jawing. It's time to get serious about your over-the-air TV future and start shopping for the most important piece of equipment: a cup holder for your reclining chair. Kidding! I speak, of course, of the antenna you need to intercept all those over-the-air TV signals whizzing overhead.

Antenna styles

The first thing you need to know about over-the-air TV antennas is that they come in three main form factors:

Location	Style	Description
Indoor	Flat	Usually a thin square, rectangle, or circle designed to be mounted in a window or similar glass surface.
Indoor	Tabletop	Usually thin, cylinder-like, and a few inches tall, with a base that enables you to sit the antenna on a table or desk.

Location	Style	Description
Outdoor	Various	Usually fairly large and designed to be mounted on a roof, in an attic, on a pole, or on the side of a building.

Determine the range

Usually, one of your first antenna decisions is the range you require. Here, *range* refers to the most distant transmission signal that the antenna can still pick up. For example, if an antenna is marketed as having a 30-mile (50-kilometer) range, it can usually pick up signals from broadcast towers up to 30 miles (50 kilometers) from your location.

REMEMBER

I wrote *usually* because the advertised range is valid only for ideal conditions. There are lots of factors that influence the strength of a signal. I go through these factors in Chapter 3, so I won't repeat them here. For your antenna shopping purposes, however, note that your real-world range for a particular antenna will almost certainly be less than what the manufacturer claims.

If you've searched for nearby stations using a tool such as TV Fool (see Chapter 3), you already have a good idea of the distance to the farthest station you want to watch. Use that value as your guide when looking for an antenna. For example, if a station you want to watch is 45 miles (72 kilometers) away, you'll want an antenna with a range of at least 50 miles (80 kilometers).

TIP

The claimed ranges of many antennas are, to say the least, far-fetched. To learn why, see the section, "Watch out for bogus or exaggerated claims," later in this chapter.

Indoor or outdoor?

An important antenna buying consideration is the choice between an indoor or an outdoor antenna. How do you make that choice? Here are some factors to bear in mind:

>> **Range:** Generally speaking, indoor antennas can pick up signals from towers up to about 50 miles (80 kilometers) away, and outdoor antennas can pick up signals from up to about 70 miles (112 kilometers) away.

>> **Cost:** Indoor antennas tend to be cheaper than outdoor ones.

>> **Appearance:** Indoor antennas are usually either so small and thin that they're barely visible once installed or have a stylish design that won't hurt your eyes. Outdoor antennas, on the other hand, tend to be quite large and ungainly and so are designed to be installed out of sight.

>> **Ease of installation:** Indoor antennas usually require minimal setup, whereas outdoor antenna installation can be arduous (depending on the existing cable connections in your home).

TIP

Some antennas are manufactured to be used indoors or outdoors. That is, the antenna offers a simple mounting system for indoor use, but the antenna is also weatherproof and comes with mounting hardware for a roof, wall, or attic.

Most people who live in an urban locale opt for an indoor antenna because it's cheaper and easier to install than an outdoor antenna, and the range is usually more than enough to pick up quite a few stations. The farther away from an urban center you live, the greater the need for an outdoor antenna.

Amplified or not?

What does an amplifier do? Here's an analogy: First, picture water coming out of a garden hose at a steady rate with the tap turned on full. If you want to increase the water pressure (perhaps to clean a stain on your deck) but the tap is on full, what can you do? That's right: You use your thumb or a finger to partially block the end of the hose, which causes the water to shoot out with greater pressure.

That's essentially what an antenna amplifier does with an incoming over-the-air TV signal: It increases — or *amplifies* — the power of that signal.

When deciding whether to get an amplified antenna, the most important fact to bear in mind is this: Amplifiers boost power only for signals that you already receive; they don't do anything for signals that are too weak to even reach your antenna. In the garden hose analogy, when you partially block the hose with your thumb, you don't get more water; instead, you boost the power, so to speak, of the existing water stream.

If you live in a rural area and the station you want is 100 miles (160 kilometers) away, an amplifier isn't going to help you get that station. Instead, you need an amplifier if the stations you do receive are weak and tend to cut out. In that case, you have a few choices:

>> **Built-in amplifier:** The amplifier is part of the antenna itself (this is common with outdoor antennas) or is attached inline with the antenna's coaxial cable.

>> **External amplifier:** The amplifier is a device that you purchase separately and connect to your antenna using coaxial cable.

>> **Distribution amplifier:** The amplifier is a coaxial splitter that also amplifies the signal to make up for the power loss that always occurs whenever you split a signal.

TECHNICAL
STUFF

When you shop for an amplifier, you'll see a lot of talk about the amplifier's gain. *Antenna gain,* or just *gain,* refers to the number of decibels (db) of power that the amplifier boosts the signal.

Note, too, that all amplifiers require electrical power, so you'll need to factor the proximity of an electrical outlet into your amplified antenna installation plans.

WARNING

Remember that an amplifier is almost always a bad idea if the over-the-air TV signals you receive are already strong because boosting strong signals can create a phenomenon called *self-oscillation* that can overwhelm your TV tuner. However, a *variable* amplifier enables you to adjust the signal boost higher or lower, so it might be an option for you.

Unidirectional or multidirectional?

I mention in Chapter 3 that the signal analysis results provided by the TV Fool website include a radar chart that shows you the direction of each broadcast tower from your location. The pattern you see in that chart determines another antenna feature:

» **Unidirectional:** The antenna essentially brings in signals from a single direction, although in practice the towers can be up to 45 or even 90 degrees apart. This antenna is also called a *directional* antenna. You want a unidirectional antenna if all or most of your available broadcast signals are clustered together, as shown in Figure 4-2.

» **Multidirectional:** The antenna brings in signals from any direction. This type of antenna is also described as *omnidirectional*. You want a multidirectional antenna if the broadcast signals in your area are scattered widely, as shown in Figure 4-3.

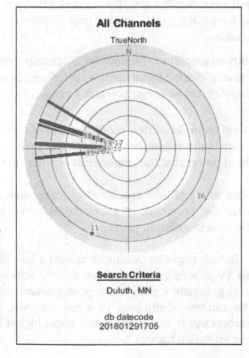

FIGURE 4-2: Get a unidirectional antenna if your incoming signals come more or less from one direction.

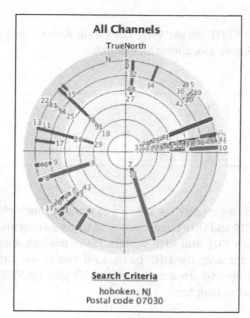

All Channels

TrueNorth

Search Criteria

hoboken, NJ
Postal code 07030

FIGURE 4-3:
Get a
multidi-
rectional
antenna if
your
incoming
signals
come
from all
over the
place.

All over-the-air TV transmissions bounce off buildings and any other objects between the tower and your antenna. Some of these bounces are redirected to your antenna. A unidirectional antenna ignores these reflected signals, but a multidirectional antenna picks them up, which can result in a common problem called *multipath distortion*.

TIP

What do you do if you live in an area where you have clusters of OTA signals coming from two different directions, but you have only a unidirectional antenna? One solution is to use an antenna rotator device, which enables you to remotely turn the antenna to face the direction you want, but rotators can play havoc with DVRs. A better solution, if you can afford it, is to buy a second unidirectional antenna, point one antenna to each signal cluster, and then use an *antenna coupler* (also called an *antenna combiner*) to join the incoming signals into a single output.

Look for VHF/UHF support

If you run a TV signal analysis report using TV Fool (www.tvfool.com; see Chapter 3), the results include a breakdown of where the available signals appear in the VHF and UHF frequency bands (see the sidebar "UHF versus VHF: Let's Talk Radio," earlier in this chapter). Figure 4-4 shows an example.

FIGURE 4-4: Available channels appear in the VHF and UHF bands.

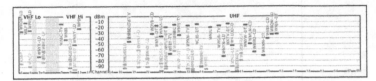

Why is this important? Because if you want to watch channels in both the VHF and UHF bands, you need to get an antenna that can pick up both VHF and UHF signals. Many modern antennas are optimized for only the UHF band, so if you're not careful, you might end up with an antenna that can't pick up VHF stations. Here's what to look for:

>> The antenna description says that it supports either VHF/UHF or dual band.

>> For VHF support, make sure the antenna supports both low VHF (lo V) and high VHF (hi V). Many antennas that have VHF hardware pick up only high-VHF channels. (Note that you need low-VHF support only if the TV Fool report indicates that you have one or more stations still broadcasting in the lower VHF frequencies.)

Watch out for bogus or exaggerated claims

Although the number of people switching to over-the-air TV has slowed in recent years, for a while it seemed like every Tom, Dick, and Harriet was cutting the cord and settling down to watch shows over-the-air. That surging popularity of over-the-air TV meant that lots of people were in the market for an HDTV antenna. And whenever you have a surging marketplace, you inevitably have scammers and unscrupulous marketers looking to take advantage.

So, to help you avoid getting scammed or duped, here are a few things to watch out for when you're researching an HDTV antenna:

>> **The antenna supports 4K or UHD.** This claim refers to the ATSC 3.0 standard that I talk about in Chapter 3, where signals will be broadcast in UHD. However, you do *not* need a special antenna to capture these signals. Any antenna that can capture run-of-the-mill over-the-air TV transmissions will also be able to pick up ATSC 3.0 transmissions.

>> **The antenna can pick up CNN, ESPN, or HBO.** Nice try, but CNN, ESPN, HBO, and the like are available only via cable or a streaming service, not over-the-air.

>> **There are no monthly fees or contract.** Well, yes, this is technically true, but there are no fees or contracts for *any* OTA TV antenna, so it's hardly something the manufacturer should be touting as a major feature.

>> **The antenna has a range of 100 miles (160 kilometers) or more.** Nope, sorry, not gonna happen. The Earth, you see, is round. That curvature means that when a broadcast tower gets beyond a certain distance from you, it falls below the horizon. That distance depends on the height of the tower, whether it's located in a high spot, and the height of your antenna. However, the maximum distance in most cases is about 70 to 75 miles (112 to 120 kilometers). Beyond that distance, there's no longer a line-of-sight between the tower and your antenna — and that means no signal. (Signals might still get to you via diffraction or tropospheric scattering, but neither of these is made possible by the antenna.)

TIP

To avoid antennas that make these bogus claims, it's generally best to stick with an antenna from a reputable manufacturer, such as AmazonBasics, Antennas Direct, Antop, Channel Master, Mohu, or Winegard.

Installing Your Antenna

After you receive your antenna, you'll no doubt be itching to get it connected to your TV or DVR. Who can blame you? Not I, dear reader, not I. However, you've just spent a bunch of time figuring out the best antenna for your needs (right?), so instead of leaping directly to the connection stage (which I cover later in "Connecting Your Antenna"), it's crucial to take time now to make sure your antenna is installed correctly. A proper installation can often make the difference between picking up just a few stations or a few dozen.

Installing an indoor antenna

The main benefit of going the indoor route for your antenna is that installation is generally a breeze:

Antenna type	Instructions
Flat antenna	Position the antenna in a window or other glass surface (such as a patio door). You can rest the antenna on a sill, or you can use two-sided tape (some antennas include pads with two-sided adhesive) to stick the antenna to the glass surface.
Tabletop antenna	Attach the antenna to its stand and then place the antenna and stand on the table or desk you want to use.

Here are a few tips to bear in mind when positioning your indoor antenna:

TIP

>> **Experiment.** You'll need to play around a bit to find the antenna position that brings in the most channels and the strongest signals. Don't be afraid to try lots of different positions to see what works and what doesn't. Remember to rescan your channels each time you change the antenna position. (See "Scanning for OTA channels," later in the chapter.)

>> **Make the position temporary, at first.** If you're installing a flat antenna, use masking tape or something similar as a temporary way of holding the antenna in place. When you

find the ideal spot, *then* use a two-sided adhesive for a long-term hold.

>> **Higher is better.** Position your antenna at the highest possible point for the best reception.

>> **Avoid large objects.** Try to position the antenna so that it's not pointing directly at a large object, such as a building or tree.

>> **Avoid metal.** If you're installing a flat antenna on glass, keep the antenna away from any metal grates or mesh that are part of the window or door.

>> **Avoid interference.** Don't position your antenna near a source of electromagnetic signals, such as your TV, a computer, an LED light bulb, or just about any electronic device.

Installing an outdoor antenna

Installing an outdoor antenna generally involves the following steps:

1. Affix the antenna to its mount.

2. Attach the mount to a surface.

The surface might be your roof, an exterior wall, a pole, a balcony railing, or an attic wall.

3. Drill a hole where you want the connection to enter your house.

4. Run some coaxial cable from the antenna to the hole you drilled in Step 3.

5. Install a coaxial cable wall plate and connect the coaxial cable to the plate.

Unless you're *very* handy around the house, I probably lost you at Step 3. If so, not to worry, because it's easy enough to hire some-one to perform all of these steps for you.

However, Steps 3 through 5 are necessary only if your home doesn't have an existing TV infrastructure. My assumption in this book, though, is that you've recently (or soon will) cut the cord with your cable company. Since you had (or have) cable TV, coaxial cables, coaxial wall plates, and other TV stuff are already installed.

You can take advantage of that existing infrastructure to make it much easier to install your outdoor antenna. All you need to do is find the main cable that enters your home and connect your antenna to that cable. That task will be either easy or hard:

>> **Easy (Option A):** Look for a coaxial extension adapter that joins two segments of that main cable. Leave the cable segment that enters your home attached to the adapter, disconnect the cable from the other end, and then screw your antenna's coaxial cable into the open end of the adapter.

>> **Easy (Option B):** Look for an endpoint of the main cable, which might be connected to a box attached to your house. Disconnect that endpoint, and then use a coaxial extension adapter to join the main cable and your antenna's coaxial cable.

>> **Hard:** If there's no extension adapter or endpoint, you need to cut the main cable and install an F-type fitting on the cut end. (There's a bit more to it than this, which is why this is the hard method. However, tons of YouTube videos can show you the full procedure.) Then use a coaxial extension adapter to join the main cable and your antenna's coaxial cable.

Here are a few tips to bear in mind when positioning your outdoor antenna:

TIP

>> **Height is your friend.** Position your antenna as high as possible to get the strongest signals with the least interference.

>> **Watch where the antenna points.** A unidirectional antenna needs to point where the transmission towers are located. But even if you have a multidirectional antenna, make sure it's not pointing directly at any large objects, such as trees or buildings.

>> **Take weather into account.** Most outdoor installs use at least one coaxial extension adapter, which can be the cause of all kinds of problems if water gets inside the adapter. To prevent that, use a coaxial extension adapter that has rubber boots on each end to protect the connections. Alternatively, you can buy separate rubber boots that you can fit over the connections.

>> **Terminate unused endpoints.** If you're reusing existing TV infrastructure, take a look around to see if any coaxial endpoints are unused. If you find any, be sure to add a terminator device to each endpoint to prevent signal leakage.

Connecting Your Antenna

Assuming you have your OTA antenna installed, your next chore is to connect the antenna to whatever device you want to use to watch over-the-air TV. That device will usually be your TV, but it could also be multiple TVs, a computer, a smartphone or tablet, or a DVR. The next few sections go through the details of each connection type. But first, you need to know how coaxial connections work.

Making coaxial connections

After you have all your equipment gathered around you, combining all those doodads into a working over-the-air TV configuration means connecting them together. Except for power, all the connections you make are coaxial, meaning they involve fastening together one or more coaxial cables as well as coaxial devices such as extension adapters, splitters, amplifiers, filters, and terminators.

The endpoint of every coaxial cable and device is called an *F connector* (or sometimes an *F-type connector*), and there are two types, as shown in the following table:

F connector	Type	Description
	Plug	A pin in the center, threads on the inside, and a rotatable nut on the outside
	Jack	A socket in the center and threads on the outside

REMEMBER In the table, the photo I use to illustrate an F connector jack is an example of a coaxial extension adapter. This device has jacks on both ends and enables you to connect two coaxial cables by inserting one cable's plug into one of the adapter's jacks, and the other cable's plug into the second of the adapter's jacks.

REMEMBER For historical reasons that are too sexist (and, I suppose, too obvious) to get into here, plug and jack F connectors are referred to distressingly often as male and female F connectors, respectively. Sigh.

Every coaxial connection involves inserting a plug F connector into a jack F connector, as follows:

1. **Line up the pin of the plug F connector with the socket of the jack F connector.**

2. **Insert the pin into the socket.**

3. **Rotate the plug F connector's outside nut (clockwise, if the plug is facing away from you) until you can't turn it any more.**

   You're connected!

Connecting an antenna to one TV

The simplest connection you can make in your OTA setup is to connect your antenna to a single TV. That connection consists of running the antenna's coaxial cable — either directly from the antenna itself or indirectly from a wall plate connected to an outdoor antenna — to the jack F connector on your TV. The jack F connector usually has one or more of the following labels:

>> Ant

>> Ant In

>> Antenna

>> Antenna In

>> Cable

>> Cable In

Figure 4-5 shows an indoor flat antenna connected to a TV's jack F connector and ready to be installed.

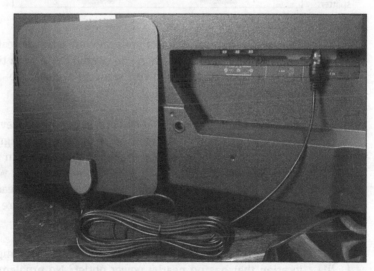

FIGURE 4-5:
Connect your antenna's coaxial cable to your TV's jack F connector.

Connecting an antenna to multiple TVs

If you want to distribute your OTA signal to two or more TVs (or other devices), you need a *coaxial splitter*. A splitter is a device that has a single jack F connector input port (usually labeled In) and two or more jack F connector output ports (usually labeled Out). Figure 4-6 shows an example.

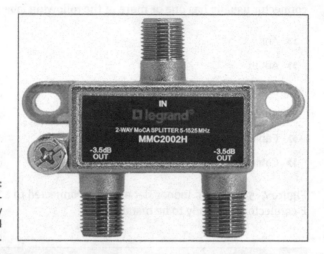

FIGURE 4-6:
A garden-variety coaxial splitter.

The idea is that you connect your antenna's coaxial cable to the splitter's input port, and then run coaxial cable from the output ports to each device that you want to receive the antenna's signals.

If you're splitting your signal and one of the output devices is far away (say, more than 50 feet, or 15 meters), the signal might degrade too much for it to be usable on that device. In that case, you might want to invest in a *distribution amplifier*, which is a splitter that boosts the signal as it goes through each of the amplifier's output ports. Distribution amplifiers are powered devices, so you'll need a nearby power outlet.

TIP

What if the perfect place to install your distribution amplifier is a location that has no nearby power outlet? No problem. Just get yourself a *power over coax* device, which plugs into a power outlet and offers two jack F connectors. Use one connector to run a coaxial cable to your TV. Use the other connector to run some coax to the distribution amplifier. That second coax cable will also power the amplifier.

Connecting an antenna to your computer

Long gone are the days when a TV was the only screen in the house. Nowadays, our faces are aglow with all kinds of screens, including computer monitors. But can you watch over-the-air TV on your computer?

The short answer is, no, you can't, because most computers are missing a crucial bit of tech: an HDTV tuner. Without that tuner, your computer doesn't have the faintest idea what it's supposed to do with an over-the-air TV signal.

Fortunately for you, the long answer is, yes, you can, as long as you augment your computer with an *external HDTV tuner*, which is a device that has a USB connector on one end and a jack F connector on the other (see Figure 4-7). Plug the device into an available USB port on your PC, run your antenna's coaxial cable to the tuner's jack F connector, and you're good to go.

FIGURE 4-7:
An HDTV tuner enables you to view over-the-air TV on your PC.

Connecting an antenna to your mobile device

If you want to watch over-the-air TV on your smartphone or tablet, you have a couple of choices:

>> If your mobile device has a USB port (as do many Android phones and tablets), look for an HDTV tuner that connects to the USB port and also offers a jack F connector to connect your antenna's coaxial cable.

>> For all other mobile devices, it's possible to broadcast your over-the-air TV signals through your Wi-Fi network, and then use an app on your mobile device to tune in to those signals. I talk about this setup in detail in the next section.

Connecting an antenna to your Wi-Fi network

If you want to distribute your over-the-air TV signals to multiple devices scattered throughout your house, what are your options? Two come to mind:

>> If your house is already wired with TV infrastructure such as coaxial cables and coaxial wall plates, you can use that setup to distribute your OTA signals. This option doesn't help if you have home areas without TV infrastructure.

>> You can run your own coaxial cables throughout the house to each place you need it. Yes, that almost certainly means drilling holes in walls and floors.

These options work in a pinch, but neither is ideal. A much neater, more flexible, and more modern solution is to take advantage of something you probably already have: a Wi-Fi network. This solution requires an HDTV tuner or DVR that can connect to your Wi-Fi network. Setting everything up generally involves these steps:

1. Connect the tuner or DVR to your antenna using a coaxial cable.

2. **Connect the tuner or DVR to your Wi-Fi router (often directly using an Ethernet cable).**

3. **Install the device app on any compatible machine.**

   Compatible machines might be an iOS or Android mobile device or a streaming device such as a Roku player or Amazon Fire TV.

4. **Use the app to locate the tuner or DVR on your Wi-Fi network.**

5. **Use the app to watch (and possible record) over-the-air TV.**

Improving antenna reception

Getting free HD channels just by installing an antenna is one of the best things about cutting the cord. Alas, those free signals lose their luster when the reception cuts in and out or falls off the digital cliff. That's just life in OTA City, right? Well, it might be if you live a really long way from the transmission towers you're trying to access. Otherwise, if your reception is problematic, try the following fixes:

>> **Look for physical interference.** Make sure your antenna's line-of-sight with any transmission tower isn't blocked by a brick wall, a building, a mountain, or tall trees.

>> **Look for electromagnetic interference.** Make sure your antenna isn't close to a TV, computer, Wi-Fi router, or another electronic device that could create interference.

>> **Add an LTE filter.** LTE cellular signals are pretty much everywhere these days, but you can eliminate them as a source of interference by installing an LTE filter. Connect the filter directly to your TV's jack F connector or to the input jack of the amplifier or splitter, if you use one of those devices.

>> **Shorten the cable.** The longer the cable, the more the signal degrades. If you're using, say, a 50-foot coaxial cable to reach a TV that's just 10 feet away, try a shorter cable.

>> **Lengthen the cable.** If you use an extension adapter to combine two shorter lengths of cable, you lose some signal through that adapter. Try a longer cable that doesn't require an extension adapter.

>> **Replace the cable.** If your setup includes an old coaxial cable, the cable might not have enough shielding. If possible, try swapping the old cable for quad shield cable.

>> **Remove the amplifier.** If your antenna includes an amplifier — or if you've added an amplifier to your configuration — try turning off or removing the amplifier.

>> **Add an amplifier.** If you've split the incoming signal, you can make up the signal loss by adding a distribution amplifier.

TIP

If you implement any of these fixes, be sure to rescan for OTA channels, as I describe in the next section.

Scanning for OTA Channels

After you have your antenna installed and connected to your TV, your final chore (not including microwaving some popcorn for the upcoming TV binging) is to convince your TV to scan for all the channels that are now available.

You might not have to do any manual scanning because many TVs are set up to automatically scan the connected coaxial cable for signals. If your TV doesn't scan automatically, however, you need to run a scan by following these general steps (the specifics of which vary depending on the make and model of your TV):

1. **On your TV's remote, press the button that takes you to the TV's setup screen.**

   The button will be labeled Menu, Options, or Setup.

2. **In the setup screen, navigate to and select the item for working with an antenna or over-the-air TV.**

   The item will be named something like Antenna, Channels, Live TV, or Broadcast.

3. **Select the item for scanning, which will be named something like Channel Scan or Channel Tuning.**

If your TV doesn't appear to have a dedicated feature for scanning OTA channels, use the Input button to select the Antenna input, which should start a scan automatically.

Your TV then proceeds to scan the incoming antenna signal for available channels and displays its progress, as shown in Figure 4-8. Note that the scan can take between 5 and 30 minutes, depending on your TV and configuration.

FIGURE 4-8: The screen shows the progress of the channel scan.

Chapter **5**

Watching Over-the-Air and Live TV

W hether you're looking to watch today's big game, this afternoon's classic soap, or tonight's prime-time drama, your antenna is (presumably) ready to pull down the channel of your choice. And that choice is what this chapter is all about. Here I finally put the OTA pedal to the TV metal and show you how to watch over-the-air programming — on not only a regular TV but also a smart TV and a streaming media device. Talk about choices!

You also explore the world of recording over-the-air shows using DVR devices and software. If you've turned your back on cable, the combination of watching and recording over-the-air shows is sure to allay any post-cord-cutting regret you might be experiencing.

Finally, if you decided not to go with an antenna for over-the-air access, this chapter also lets you investigate the world of live TV streamed over the Internet. A fistful of live TV streaming services are available at every price range, so there's bound to be one that's right for you. Let's see.

Watching Over-the-Air TV Using a Regular TV

Assuming you have your HDTV antenna connected to your TV's antenna jack (which might be labeled Antenna, Antenna In, Ant, or Cable; see Chapter 4), you're ready to watch over-the-air TV. The next few sections explain the details.

Changing the TV input source

If you pull out your TV and take a look at the back (and sometimes also the side), you'll see a panel (or two) bristling with ports and jacks and connectors of various sizes and shapes (see Figure 5-1).

On any modern TV, you're likely to see one or more of the following connector types for video input:

>> **Antenna:** A jack F connector for the coaxial cable from your HDTV antenna

>> **HDMI:** A high-definition media input connector for input from a media gadget such as a streaming device, DVR, Blu-ray player, or computer

>> **Component:** Red, green, and blue RCA connectors for video input

>> **Composite:** A yellow RCA connector for video input

>> **S-Video:** A round connector for Separate Video input

>> **Computer:** A DVI or VGA connector for input from a computer monitor (although the vast majority of modern monitors connect via HDMI)

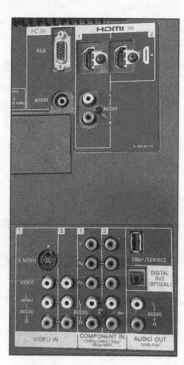

FIGURE 5-1: The back of a typical modern TV is festooned with connectors.

So, how do you tell your TV which of these connectors has the content you want to view onscreen? You need to change the *input source*, which is the connector type that the TV uses to display incoming content. If you want to watch over-the-air TV, you switch to the antenna input source; if you want to watch shows using a streaming media device or DVR, you switch to the HDMI input source.

The method you use to change the input source depends on your TV. The most common case is to use the TV's remote. Look for a button named Input (or sometimes Source), as shown in Figure 5-2, and then use one of the following methods to switch to a different input source:

>> **Each time you press Input, the TV displays the next input source:** Keep pressing Input until you get to the input source you're looking for.

>> **When you press Input, a menu of input sources appears:** Use the remote's navigation ring or navigation buttons to select the source you want to view.

FIGURE 5-2:
For most
TVs, you
use the
remote's
Input
button
to change
the input
source.

TIP

If you don't see a button named Input (or Source) on your remote or you don't have a remote, look on the TV itself, which should have an Input (or Source) button.

Watching over-the-air TV

Checking out what's currently being broadcast over-the-air is so straightforward that I can describe it in a mere two steps:

1. **Change your TV's input source to Antenna.**

The name of this input source varies depending on the TV, so if you don't see Antenna, look for TV.

2. **Use the TV remote's Channel Up and Channel Down buttons to surf the available channels.**

Alternatively, if you know a station's virtual channel number (see Chapter 3), you can use the remote's numeric keypad to enter the channel number.

When you enter the virtual channel number, remember to include the decimal. For example, if the virtual channel number is 4.1, you need to press 4, a period (or a hyphen on some TVs), and then 1.

Where's the TV guide?

The basic over-the-air TV watching experience involves surfing up and down through the available channels until you find something entertaining, interesting, or diverting enough to put down the remote.

Channel surfing is a classic way to watch TV, but it gets old in a hurry. What's the solution? Right: Some sort of guide to what's on now and what's coming up. Fortunately, plenty of free online services provide a TV listing based on your location. Here are a few to check out:

>> **NoCable:** Provides over-the-air TV listings for any address in the United States, as www.nocable.com/tv-guide/

>> **On TV Tonight:** Provides OTA TV listings (for morning and afternoon, as well as night) for cities in the United States, Canada, United Kingdom, Australia, and Ireland at www.ontvtonight.com/

>> **TitanTV:** Provides over-the-air TV listings for any ZIP code in the United States at www.titantv.com/

>> **TV Passport:** Provides over-the-air TV listings for cities and ZIP or postal codes in the United States and Canada at www.titantv.com/

Quite a few mobile device apps can also provide local TV listings. The best of these is the TV Guide app, which is available for both iOS and Android.

Recording over-the-air TV

It will probably take you less than ten minutes to stumble upon the biggest problem with watching TV shows over-the-air: commercials, *so* many commercials. In your cable days, you likely solved that problem by recording your favorite shows and then skipping over the commercials during playback. Too bad you can't do that with over-the-air TV, right?

Wrong! Devices and software are available that let you set up a roll-your-own DVR setup. The next couple of sections run through the most popular options.

Recording over-the-air TV for free

As I describe in the next section, the best over-the-air DVR experience requires a subscription, which gives you sweet features such as scheduled recordings and recording only new episodes of a show.

However, if you're not fussy about DVR features — or your budget is tight — there are ways to record over-the-air TV without paying a monthly subscription. These free DVR solutions are usually restrictive — for example, they might support only manual recordings or your recordings might be playable only on a single device — but, hey, whaddya want for nothing?

Here are some free DVR setups to check out:

>> **Tablo:** Most people use the Tablo Dual or Tablo Quad DVR with a subscription (see the next section) to get its abundant features, but these Tablo devices also offer basic recording capabilities without a subscription. How basic? Well, you get just one day of TV guide data; you get no program info, and you can't record by series. I show you how to configure a Tablo DVR a bit later in this chapter (see "Setting up a Tablo DVR").

>> **Fire TV Recast:** If your home is a dedicated Amazon Fire TV shop, Fire TV Recast is for you. Recast is an over-the-air TV that connects to your home's Wi-Fi network, where it can then be accessed by any Fire TV device. Fire TV Recast includes either two tuners and 500GB of internal storage, or four tuners and 1TB of storage.

As I write this, Fire TV Recast is available only in the United States.

REMEMBER

>> **Live Channels:** This is a Google-created Android app, so it works on any Android TV device. The app is free to

download and requires no subscription. If your Android TV box doesn't come with a built-in tuner (such as you get with AirTV and Channel Master Stream+), you need to buy a tuner (such as the HAUPPAUGE WinTV-DualHD tuner). You'll also need a USB hard drive to store the recordings.

>> **AirTV Player:** This option comes with free DVR capabilities via either the AirTV app or the Sling TV app. It has a tuner, but you need to attach a USB hard drive before you can record anything.

>> **Digital converter box:** This inexpensive device is designed to convert digital over-the-air signals so they can be displayed on analog TVs. However, they work perfectly well with any TV, include a tuner, and also offer extremely basic (and extremely free) recording capabilities.

Recording over-the-air TV with a subscription

If you want the nicest DVR experience for your over-the-air shows, you need to shell out a few dollars a month for a DVR subscription service. Those dollars get you features such as a multi-day TV guide (usually 14 days' worth of listings); scheduled and manual recordings; series info; series-based recording; adjustable recording start and stop times; channel filters (such as showing only movies or sports); and automatic commercial skipping.

Here are three subscription-based DVRs to consider:

>> **Tablo:** The Table Dual (two tuners) and Table Quad (four tuners) are whole-home DVRs, meaning they connect to your Wi-Fi router and then other devices — such as a streaming media device, smart TV, mobile device, or computer — access the DVR over Wi-Fi using the Tablo app. Most Tablo DVRs require you to add a hard drive for storing your recordings. Basic recording is free (see the preceding section), but you'll need a subscription ($4.99 per month) to unlock all the DVR features. See the next section for the details on how to set up a Tablo DVR.

REMEMBER

The more tuners you get, the more options you have when it comes to simultaneously watching live TV and recording shows. For example, with two tuners, you can simultaneously watch one show and record another or record two shows; with four tunes, you can watch and record up to four shows at once.

>> **Channels:** You install the Channels software on a computer, connect an HDTV tuner to the same computer, get a Channels Plus subscription ($8 per month), and then start watching and recording over-the-air TV. For more details, surf to https://getchannels.com/plus/.

>> **Plex:** You install Plex Media Server on your computer, add a compatible tuner, and then subscribe to a Plex Pass ($4 per month), which enables you to watch and record over-the-air TV. You can access recordings on any networked device. For more info, see www.plex.tv/tv/.

Setting up a Tablo DVR

Tablo DVRs are by far the most popular option for cord cutters who are doing the over-the-air TV thing. If you decide to go the Tablo route, here are the general steps to follow to set up most Table DVRs:

1. Connect your HDTV antenna's coaxial cable to the Antenna jack F connector on the back of the Tablo.

2. If your Tablo doesn't come with a built-in hard drive, connect an external USB hard drive to the USB port on the back of the Tablo.

   Some Tablo models also accept an internal SATA hard drive. If you're using this method, remove the screw on the bottom of the Tablo that holds the cover for the SATA compartment in place, insert the drive into the SATA connector, and then reattach the SATA compartment cover.

3. Access the Tablo app with the device you want to use to set up the Tablo:

   • If you're using a PC or Mac, browse to the Tablo web app at http://my.tablotv.com/.

   • If you're using an iOS or Android mobile device, install the Tablo app.

4. **Connect your Tablo to your Wi-Fi network:**

- If your antenna and your Wi-Fi router are close to each other, attach one end of the supplied Ethernet cable to the Tablo's Ethernet port and the other end of the cable to any available Ethernet port on your router. Skip to Step 7.

- If your antenna and Wi-Fi router are too far apart for a direct connection, you need to use Wi-Fi, so select the app's Add via WiFi command. Your Tablo sets up its own wireless network with a name along the lines of Tablo_1234.

5. **Use your computer or mobile device to connect to the Tablo wireless network.**

6. **Use the Tablo app to select your home Wi-Fi network and enter the network password.**

This step connects your Tablo to your network.

7. **Follow the instructions provided by the Tablo DVR Setup Wizard.**

8. **If you also use a smart TV or streaming media device, use that device to install the Tablo app so that you can access Tablo content on the device.**

Watching Over-the-Air TV Using a Smart TV

If you have a smart TV — such as an Amazon Fire TV Edition set, a Roku TV, an Android TV, or a Samsung Smart TV — in many cases you can watch over-the-air TV just by connecting your HDTV antenna to the device.

Most smart TVs have the input source selection built right into the TV's interface. For example, on Fire TV Edition televisions, the Home screen (as well as the Live screen) offers an Inputs section, as shown in Figure 5-3. You use this section to choose the input source you want to view. For over-the-air TV, you'd select the Antenna input source.

FIGURE 5-3:
With smart TVs, you use the TV interface to select the input source.

Select the Antenna input to watch over-the-air TV

After you select the input source for your antenna, you usually see what's playing on the current OTA channel. The interface typically provides a button to display a channel guide like the one shown in Figure 5-4.

FIGURE 5-4:
Smart TVs offer a channel guide.

Watching Over-the-Air TV Using a Streaming Device

If you have a media streaming device such as Roku, Fire TV, or Chromecast, can you use that device to watch over-the-air TV? Not directly. Instead, you need either a DVR or an HDTV tuner that will send your antenna's OTA signals via Wi-Fi to the streaming device, and then you use the DVR or tuner app to watch the signal.

If you want to use a DVR, any of the following will do the trick (see my descriptions of these DVRs earlier in the "Recording over-the-air TV" section):

>> Tablo

>> Fire TV Recast

>> AirTV

If you just want to use an HDTV tuner, here are a couple to investigate:

>> **HDHomeRun:** Connect your antenna's coaxial cable to the back of this device. Then connect the HDHomeRun to your Wi-Fi network, either directly using an Ethernet cable or by using a temporary wireless network.

>> **ClearStream TV:** Connect your antenna's coaxial cable to the jack F connector on this device, and then use the ClearStream TV app to connect the tuner to your Wi-Fi network.

Look, Ma, No Antenna! Streaming Live TV

In the three chapters in Part 2 (Chapters 3, 4, and 5), I make a big assumption: Your post-cable life includes over-the-air TV brought into your home using an HDTV antenna. That's an attractive proposition because most folks can pick up one or two dozen stations broadcasting in crystal-clear HD, all for the price of an antenna and some coaxial cable. Sweet!

But over-the-air TV slurped up by an antenna isn't for everyone:

>> Perhaps you live in an area that has very few over-the-air broadcast signals.

>> Perhaps the only way you can get a reasonable number of channels is to use an outdoor antenna, and you don't want to bother with the effort or expense of installing one.

>> Perhaps you really want DVR capabilities, but you find the do-it-yourself route of installing and configuring a DVR such as a Tablo or Fire TV Recast too complicated.

Whatever the reason, does it mean you have to give up on watching or recording live TV? Not even close. Tons of streaming services offer *live TV*, meaning local channels broadcast in real time. You use the service's app to access the live TV feed. In most cases, you can use the app on any of the following device types:

>> A desktop or notebook computer (PC or Mac)

>> iOS and Android mobile devices

>> Gaming consoles, particularly Xbox and PlayStation

>> Streaming media devices, such as Roku, Fire TV, Apple TV, and Chromecast

>> Smart TVs, such as Fire TV Edition sets, Roku TV, Android TV, and Samsung TV

Many live TV streaming services provide access to dozens of channels and offer some sort of *cloud DVR*, which refers to a digital video recorder that stores recordings online. The services range from free to eyebrow-raisingly expensive, but there's bound to be something that fits what you're looking for. To find out, check out the most popular services, listed here:

>> **AT&T TV Now:** $69.99. Offers more than 65 channels of live TV. Includes 20 hours of cloud DVR storage. You can get 500 hours of cloud DVR storage for an extra $10 per month. Go to www.att.com/tv/.

>> **Hulu + Live TV:** $64.99. Gives you access to more than 65 live channels (plus the rest of the massive Hulu library). A cloud DVR is included, but you can upgrade to the Enhanced Cloud DVR for an extra $9.99 per month. Visit www.hulu.com/live-tv.

>> **fuboTV:** $64.99 to $79.99. Offers more than 100 live channels with an emphasis on sports. A cloud DVR is included. See www.fubo.tv/.

>> **Locast:** Free. A not-for-profit that offers live, local broadcast TV in selected US cities (with more cities added somewhat regularly). For details, go to www.locast.org/.

>> **Paramount Plus:** $5.99 or $9.99. Gives you live access to CBS programming. The $9.99 option is commercial-free. See www.paramountplus.com/.

>> **Philo:** $20. Offers more than 60 channels and unlimited cloud recording. Visit www.philo.com/.

>> **Pluto TV:** Free. Offers 250 or so channels, many of which are broadcast live. Go to https://pluto.tv/live-tv/.

>> **Sling TV:** $35. Provides two different packages — one with about 45 news and entertainment channels, and the other with about 30 sports and family channels — for the same price. 50 hours of cloud DVR storage is included, but you can get 200 hours for an extra $5 per month. Visit www.sling.com/.

>> **YouTube TV:** $64.99. Offers more than 85 channels of live TV and comes with unlimited DVR cloud storage. Visit https://tv.youtube.com/.

3

Cable-Free Viewing with Streaming Services

IN THIS CHAPTER

» **Choosing a TV for media streaming**

» **Learning the difference between a set-top box and a dongle**

» **Choosing between a set-top box and a dongle**

» **Checking out smart TVs**

» **Setting up your streaming hardware**

Chapter 6

The Hardware You Need for Streaming

B y necessity, streaming media over the Internet is a some-hardware-required exercise. At the very least, you need a screen to watch the incoming media stream. And that screen might be all you need for hardware if you're watching the media on a mobile device using a cellular connection. If that scenario doesn't apply to you (and it almost certainly doesn't), your hardware investment will be more substantial. I'm talking a TV or smart TV; a streaming media device, such as a set-top box or a dongle; perhaps a digital video recorder doohickeys I talk about in Chapter 5. You might also need to beef up your Internet-access hardware and your Wi-Fi network equipment. Then, of course, you have to wire everything together so that you and yours can actually watch stuff.

Fortunately, getting and setting up your streaming hardware is nowhere near as daunting as it might seem. In this chapter, you explore all the hardware possibilities associated with streaming media, from smart TVs to streaming media devices and beyond. (Internet and Wi-Fi devices are the subject of Chapter 7.) I take you through what's available for each type of hardware, give you tips for deciding what to get, and show you how to put it all together.

Let's Talk TVs

I assume that since you once had a cable overlord, you still have the television you used to watch cable programming. If so, and you're happy with your set, there's probably no reason to hang out in this section any longer.

Why "probably"? Well, since you've gone to the trouble to cut your ties with the cable company, you owe it to yourself to create the best post-cable experience you can (within your budget constraints, of course). You might think that just means getting the most suitable streaming media player, but before getting to that, take a closer look at your existing TV. Is it *really* going to give you the best experience with streaming media? Is it the right size? Does it have the optimum resolution? Does it have the ports you need?

Good questions, all, and the next few sections provide the answers.

Screen size

You might think "the bigger the better" is the only rule you need when it comes to choosing a screen size for watching streaming media. You wish! Sure, a 75-inch set would be awesome, but if you're sitting four or five feet from that behemoth, your eyes will give out before you've finished your popcorn.

You'll get the best media streaming experience if you tailor your TV's screen size to the room where the TV will reside. In particular, you need to match the screen size to the distance you'll sit from the screen while you're watching. In general, the farther away you sit, the bigger the screen you can rock. Here's a procedure that will help you decide:

1. Measure the distance, in feet, between where you sit to watch TV and where the TV will be set up.

2. Multiply the distance in Step 1 by 12 to convert it to inches.

3. Multiply the distance in inches by 2/3 (or 0.67). The result is the maximum screen size you should buy.

Here are some examples:

Distance (feet)	Distance (inches)	Maximum screen size (inches)
10	120	80
9	108	72
8	96	64
7	84	56
6	72	48
5	60	40

Resolution

The *resolution* of the TV refers to the number of pixels it uses to display the picture. As I explain in Chapter 3, the more pixels (that is, the higher the resolution), the sharper the image. These days, you should consider three main formats when looking at TVs for streaming media:

Format	Resolution	Also known as
HD or HDTV	1920 x 1080	1080p
UHD or UHDTV	3840 x 2160	4K
UHD-2	7630 x 4320	8K

First, although 8K TVs are available, you won't find any streaming media (or streaming devices) that support 8K, so you can rule out that format for now. (Let's talk again in a few years.) That leaves HD and 4K to consider:

>> **Your TV and your streaming device should support the same format.** You're over-buying if you purchase a 4K TV

when your streaming device supports only HD. Conversely, you're under-buying if you purchase an HDTV when your streaming device supports 4K.

» **The larger the screen size, the greater the need for 4K.** On relatively small screen sizes — say, 42 inches or less — the HD format looks amazing. But as the screen size increases, those pixels have to cover a bigger area, so the picture quality diminishes. Consider 4K if you're buying a TV with a screen size of 50 inches or more. However . . .

» **The closer you sit to the TV, the greater the benefit of** only **4K.** Human eyes (even ones with 20/20 vision) can resolve so much detail on the screen, and the farther away you are from the screen, the less detail your eyes can detect. Depending on the screen size, you need to be within three to five feet of a 4K screen to notice the extra detail. If, like most people, you sit eight to ten feet away from the screen, don't bother with 4K because you won't be able to resolve the extra pixels (unless you get an 80-inch screen!).

HDMI ports

Almost all streaming devices and digital video recorders connect to a TV using an HDMI cable. That means you need a TV that has one or more (almost certainly more) HDMI ports. Really old TVs don't have HDMI ports, so if your TV is that old, it's definitely time to upgrade.

On most TVs, the HDMI port is labeled *HDMI.* If your TV has multiple HDMI ports (as most modern TVs do), the ports are usually labeled *HDMI 1, HDMI 2,* and so on, as shown in Figure 6-1. Newer TVs usually have all their HDMI ports on one side of the TV's back panel (refer to Figure 6-1), while on older TVs it's common to have one HDMI port on the bottom of the TV's back panel and a second HDMI port on the side of the back panel.

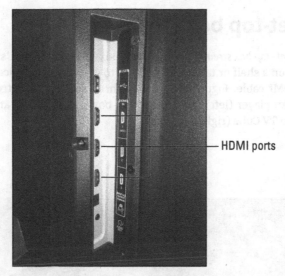

FIGURE 6-1:
Modern
TVs have
all their
HDMI
ports
together
on the
back
panel.

HDMI ports

When you're buying HDMI cables, don't go cheap, if possible. Higher quality HDMI cables can make a big difference in the quality of the image you see onscreen. For best results, buy cables that support HDMI 2.0 or, even better, HDMI 2.1.

TIP

Streamers: Set-Top Box or Dongle?

Streaming media players — often just called *streamers* by folks in-the-know — come in all shapes and sizes. However, you can simplify things by understanding that almost all streamers fall into one of the following two categories: set-top box or dongle. The next few sections examine each type.

A third category of media player is the streaming soundbar. A *streaming soundbar* is a device that combines a streaming media player with audio hardware, usually including speakers and sub-woofer. Examples include Amazon Fire TV Edition soundbars from Nebula and TCL, and the Roku Streambar.

REMEMBER

Set-top box streamers

A *set-top box streaming player* is a box-like device that's meant to sit on a shelf or table and connect to your display device using an HDMI cable. Figure 6-2 shows three set-top box streamers: a Roku player (left), an Android TV box (middle), and an Amazon Fire TV Cube (right).

FIGURE 6-2: Three examples of set-top box streaming players: Roku player (left), an Android TV box (middle), and an Amazon Fire TV Cube (right).

Set-top box streamers have a back panel that includes an HDMI port and one or more other ports for connecting devices, as shown in Figure 6-3.

FIGURE 6-3: Set-top box players have multiple input ports.

Dongle streamers

In computing lingo, a *dongle* refers to any device that plugs directly (that is, without a cable) into a port on another device, such as a computer or TV. So a *dongle streaming player* — in some cases, also called a *streaming stick* — is a device that connects directly to an HDMI port on a TV or display to provide streaming media services. Figure 6-4 shows three dongle streamers: a Roku (left), a Google Chromecast (middle), and an Amazon Fire TV (right).

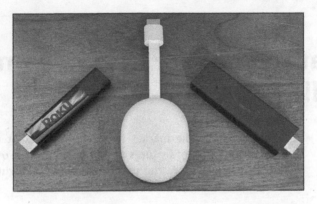

Set-top box streamer versus dongle streamer

Which type of streamer should you buy? The answer depends on the features you want and how much you want to spend. In general, set-top boxes

>> Are larger than dongles.

>> Require access to a power outlet (dongles get their power from the HDMI port to which they're connected).

>> Are a bit messier than dongles to use because they take up more room and require both an HDMI cable connection to the TV or display and a power cord connection to an outlet.

>> Usually offer more features than dongles because they can fit extra electronic wizardry into their more expansive interiors.

>> Are faster than dongles because one of the internal features you get in a set-top box is a larger and faster microprocessor.

>> Offer multiple ports, such as one or more USB ports, an audio output jack, and an Ethernet port.

>> Tend to be more expensive than dongles, which isn't surprising given the extra features and hardware.

What to Look for in a Streaming Media Player

Whether you're in the market for a set-top box or a dongle, there's a wide range of features to furrow your brow over. Here, listed alphabetically, are the main features you might want to consider when doing your research:

>> **4K support:** Gives you the best image quality when streaming media if you also have a 4K TV.

>> **AirPlay support:** Enables you to beam media from your iPhone, iPad, or Mac directly to the streamer.

>> **Audio output jack (for headphones):** Enables you to watch streaming media without bothering anyone within earshot.

>> **Bluetooth support:** Enables you to wirelessly connect a Bluetooth device to the streamer and then either access the media via the Bluetooth device (for example, audio via Bluetooth headphones) or play Bluetooth device media via the streamer.

>> **Compatibility with your existing stuff:** Gives you the easiest setup and the most reliable performance. Streamers from Amazon, Apple, and Google tend to work best with their own devices. For example, Amazon Fire TV

streamers connect seamlessly with Amazon Echo smart speakers; Apple TV devices work best with Macs, iPhones, and iPads; and Google Chromecasts connect well with Android mobile devices. If your house is an Amazon, Apple, or Google shop, consider a streamer from the same company for best compatibility.

>> **Ethernet port:** Enables you to connect the streamer directly to your Wi-Fi router for better performance. See the sidebar titled "Wired streaming," a bit later in this chapter.

>> **High dynamic range (HDR):** Gives you the best picture quality. Note that you need a TV that also supports HDR.

>> **Input ports:** Ensures that you have enough ports if you plan on expanding your system with an external USB hard drive (usually used for DVR storage) or some other external device.

>> **Remote features:** Ensures that you select a remote that matches your needs. Some streamer remotes are extremely simple and offer just a few buttons for navigating and playback. Other streamers take the opposite tack and offer remotes bristling with buttons for controlling not only the streamer but other devices such as your TV.

>> **Voice control:** Enables you to use voice commands to navigate the streamer interface, search for content, and play media:

- *Amazon Fire TV:* Works with Amazon's Alexa voice assistant. Depending on the device, it will have Alexa built in, come with an Alexa voice remote (see Figure 6-5), or connect to an Alexa-enabled device such as an Amazon Echo. You can also use the Alexa mobile app to send voice commands to your connected Fire TV device.

- *Apple TV:* Works with Apple's Siri voice assistant. The Apple TV Siri Remote comes with a Siri button (microphone icon) that you press and hold to issue voice commands.

- *Google Chromecast:* Works with Google Assistant for voice control. Depending on the device, you can access Google Assistant via the Chromecast voice remote, a

Google Assistant-powered smart speaker, or the Google Assistant app on a mobile device.

- *Roku:* The Roku Voice remote (which comes with players such as the Roku Ultra and the Roku Streambar but is also available separately) lets you use voice to control most Roku streaming devices.

FIGURE 6-5: The Alexa voice remote includes a voice button for sending voice commands to a Fire TV device.

Voice

TECHNICAL STUFF

WIRED STREAMING

One of the great conveniences of modern streaming players is that you can connect them to your Wi-Fi network, which enables the player to download media from anywhere in your house (as long as the device isn't too distant from your router). That convenience is great, but Wi-Fi has its problems: It can be slow, performance can tank if lots of other people are using the network, and other electromagnetic devices can interfere with the signal. If you're having any of these problems, one slightly out-there solution is to connect the streaming player directly to the router. If your streamer has an Ethernet port, you can run an Ethernet cable from that port to any Ethernet port on your router. Just like that, you get a faster connection speed, no performance lag, and no interference. Win-win-win!

What's the catch? Well, there are two. First, your streaming player and your router should be in the same room for the easiest cable connection. Yep, you can run Ethernet cable through and within walls, but that might be overly ambitious. Second, not every streaming player comes with an Ethernet port. As this book went to press, the following major streamers were Ethernet-friendly:

- Android TV Box

- Apple TV 4K

- NVIDIA Shield

- Roku Ultra

Adapters are available that enable you to connect a non-Ethernet player directly to your router via Ethernet:

- **Amazon Ethernet Adapter for Amazon Fire TV Devices:** Connects to the USB port on a Fire TV device. You then connect an Ethernet cable to the port on the adapter.

- **Ethernet Adapter for Chromecast with Google TV:** This power adapter connects via USB to Chromecast. You then connect an Ethernet cable to the port on the adapter.

- **Third-party Ethernet adapters:** You can find tons of third-party adapters that include an Ethernet port and connect to any streaming player that has a USB port.

Smart TVs for Streaming Media

An increasingly popular way to do the streaming media thing is to smush together a TV and a streaming device into a single gadget called a *smart TV*. It's *smart* because it has computer hardware that runs essentially the same software as a streaming set-top box or dongle. This combination of hardware and software means that as soon as you turn on the set, you see an interface for your streaming apps and shows, as shown in Figure 6-6.

FIGURE 6-6: Turn on a smart TV and you see your streaming apps and shows right away.

Why bother with a smart TV when a streaming player will do the same job? Here are a few reasons:

>> **You're buying a TV anyway.** If you're in the market for a new TV, consider a smart TV because the quality and features can be just as good as a regular TV and smart TVs don't cost much more.

>> **The hardware setup is easier.** Because the streaming player is built into the set, you don't have to worry about connecting devices together.

>> **You get more expansion options.** A smart TV has more connectors than any streaming player, which means you have more options for expanding your entertainment system for audio, video, and gaming.

>> **You use one remote.** Instead of separate remotes for your TV and streaming player, you get everything in a single remote.

>> **OTA access is easier.** If you have an antenna connected to your smart TV, most interfaces also include a way to view your over-the-air TV channels directly (that is, without having to switch inputs).

Here are some smart TV types to consider:

>> **Amazon Fire TV Edition:** Runs a version of Amazon's Fire TV software that includes extra TV-related features. Fire TV Edition sets are available from Insignia and Toshiba.

>> **Android TV:** Runs a version of the Android operating system that includes features for streaming media. Android TV sets are available from Hisense, Sceptre, and Sony.

>> **LG:** Offers streaming apps via LG's web OS operating system.

>> **Roku TV:** Runs a version of Roku's streaming software. Roku TV sets are available from Hisense, RCA, and TCL.

>> **Samsung:** Offers streaming apps via Samsung's Tizen operating system.

Setting Up Your Hardware

If you're a certain age, you may remember when devices were advertised as being plug-and-play, which meant, at least in theory, that you simply connected the device and it would configure itself automatically, meaning you could then play with the device (whatever that meant) after a minute or two. (In practice, such devices were better described as plug-and-pray.)

I'm sorry to report that your streaming player does *not* fall under the plug-and-play rubric. Instead, after you plug in your device, you must run through a nontrivial setup process before you can play with it. That process includes crucial steps such as plugging in the device, connecting to your Wi-Fi network, and signing in to your account. Lucky for you, the entire process usually takes only a few minutes.

Connecting a set-top player

If you have a set-top streaming player, you need to position the device optimally, connect the device to your TV, and then trudge through the setup procedure. The next few sections explain all.

Positioning the set-top box

After you've liberated your streaming player from its packaging, one obvious question arises: Where the heck do you put it? Somewhere near your TV seems like the obvious answer, but choosing the best location is a bit more complicated. Here are some things to consider:

>> Your streaming player requires full-time power, so make sure the device is close to an outlet.

>> Your streaming player connects to an HDMI port on your TV, so make sure that the player's HDMI cable is close enough to reach the TV.

>> The streaming player must be within range of your Wi-Fi network.

>> Don't store the streaming player inside a cabinet or other enclosed location.

>> If the streaming player can recognize voice commands, TV or sound system speakers can befuddle the streaming player's built-in microphone. So make sure all speakers are at least 1 to 2 feet away from your streaming player. Also, make sure the device is close enough that you can give your voice commands without having to yell. Depending on the ambient noise in your environment, this usually means being within 15 to 20 feet of the device.

Connecting the set-top box to your TV

Your streaming player connects to your TV's HDMI port (see Figure 6-7), which on most TVs is labeled HDMI (or HDMI 1, HDMI 2, and so on).

FIGURE 6-7:
Use an HDMI cable to connect your streaming player to your TV.

With your streaming player connected to your TV, grab the power cable that came with your streaming player. Connect one end of the power cable to the power port on the back of the streaming player and plug the other end of the cable into a power outlet.

Turn on your TV and change the input source (as I describe in Chapter 5) to your streaming player's HDMI connection.

Connecting a streaming stick

Your streaming stick connects to an HDMI port on your TV. I mentioned earlier that HDMI ports reside either on a side panel of the TV or on both a side panel and the back panel, as shown in Figure 6-8.

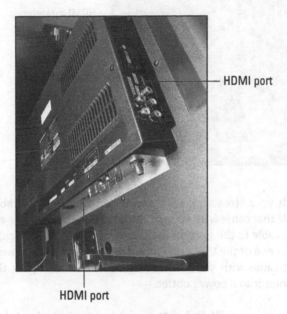

HDMI port

FIGURE 6-8: Older TVs often have their HDMI ports in multiple locations on the back panel.

HDMI port

The location of the HDMI port is important because the length of the streaming stick often means there isn't room between a bottom HDMI port and whatever surface the TV is sitting on for the streaming stick to fit. If that's the case for you, here are three possible solutions:

» Plug the streaming stick into a side HDMI port, if one is available.

>> Mount the TV on the wall (which gives the streaming stick plenty of room because there's no longer a surface immediately under the TV).

>> Use an HDMI extender cable. Insert the smaller end of the extender cable into the HDMI port on your TV, and then connect your streaming stick to the larger end of the extender cable (see Figure 6-9).

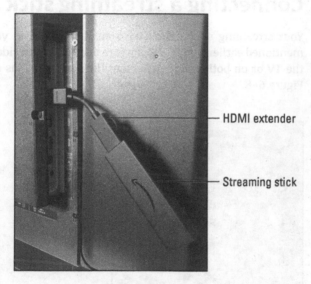

HDMI extender

Streaming stick

FIGURE 6-9:
A streaming stick with an HDMI extender cable.

With your streaming stick connected to your TV, grab the USB cable that came with your streaming stick. Connect one end of the USB cable to the port on the side of the streaming stick, plug the other end of the USB cable into the USB port on the power adapter that came with your streaming stick, and then plug the power adapter into a power outlet.

Turn on your TV and change the input source (as I describe in Chapter 5) to your streaming stick's HDMI connection.

Configuring your streaming player

When you change the input source (see Chapter 5) to the stream-ing player's HDMI port on your TV, the streaming player starts up for the first time and takes you through its setup process. This process varies from player to player, but usually includes some or all of the following tasks:

>> Connecting the streaming player to your Wi-Fi network

>> Creating or signing in to your account (such as your Amazon account if you're setting up a Fire TV device)

>> Configuring the player's initial settings

>> Connecting to the streaming player's remote

When you're finished, you're almost ready to dive into the world of streaming. First, however, you might want to see if your Wi-Fi network and Internet access are up to the task. If so, Chapter 7 is the place to be.

IN THIS CHAPTER

» Learning what to look for in a new Wi-Fi router

» Extending your wireless network to avoid dead zones

» Checking that you have enough Internet bandwidth

» Reducing streaming data usage

» Making sure your download speeds are stream-worthy

Chapter **7**

Getting Your Internet Access Ready for Streaming

One of the nice things you can say about having tradi-tional cable TV service is that it's a set-it-and-forget-it experience. That is, once you have your cable connected to the set-top box and the set-top box connected to your TV, you're good to go: Just turn on the television and start clicking through channels. With cable, there's no such thing as a bad connection because the data comes to you via a dedicated, well, *cable.* Sure, technical problems arise from time to time, but you mostly get solid service without any fuss.

Watching streaming media isn't as straightforward because that media comes to you via the Internet and then is broadcast through your home via Wi-Fi. If your Internet connection or your wireless network are slow or flaky, your streaming adventures will suffer big time.

So, all the more reason to make sure you have the Internet access and Wi-Fi network that will support your nascent streaming habit. In this chapter, you delve into the devices, plans, and architectures that should be part of any modern-day streamer's basic home configuration. If your current Internet and wireless access just isn't good enough, you learn everything you need to know to turn your experience from streaming nightmare to streaming nirvana.

Investing in a New Router

When you get into streaming media, it's easy to forget about the hardware that underlies everything. After all, after you connect a player or start your smart TV, all the streaming seems to happen right there on your screen as you navigate from app to app and show to show. If you think about where all that content comes from, it's with a vague nod to an amorphous cloud out there somewhere on the other end of the Internet's tubes.

However, for streaming media to get from the cloud out there to your TV in here, the data has to pass through what is arguably the most important — and certainly the most used — gadget in your house: your Wi-Fi router. That device is the beating heart of your wireless network. It's the workhorse through which you surf, email, chat, meet, post, and perform all the other verbs that are the hallmarks of online life.

Ah, but now you've added a new verb to that collection: *stream*. Any old bargain-basement router can surf and email and the rest of it, but asking a cheap router to stream is asking for trouble. If you're using such a router, it might be time to consider an upgrade that offers the best features for streaming. What are those features? Here's a summary of what to look for:

>> **Bands:** All modern routers support data transfers over two frequencies: 2.4 GHz and 5 GHz. If you see a router that supports only 2.4 GHz, walk away.

>> **Speeds:** Get the highest speeds you can afford, which should mean at least 500 Mbps on the 2.4 GHz band and at least 1 Gbps on the 5 GHz band.

>> **Tri-band:** This type of router offers three radio bands, usually one at 2.4 GHz and two at 5 GHz, which means you can configure the router to balance the load between bands for optimum performance, especially for streaming media.

>> **Intelligent band steering:** This feature enables the router to automatically choose the best (the fastest or least congested) band available for the data it's receiving.

>> **Quality of Service (QoS):** This feature enables you to choose an application or device or both that should receive priority access to the network (see Figure 7-1).

>> **Wi-Fi 6:** This is the latest version of Wi-Fi (also known as 802.11ax), which promises significantly faster speeds than Wi-Fi 5 (802.11ac). However, you need devices that support Wi-Fi 6 to take advantage of these higher speeds. If you don't have such devices yet, you can still go for a Wi-Fi 6 router now because it will be compatible with Wi-Fi 5.

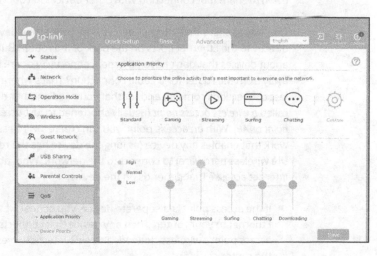

FIGURE 7-1: QoS features let you to prioritize network applications or devices or both.

MODEM? ROUTER? ACCESS POINT? WHAT'S THE DIFFERENCE?

TECHNICAL STUFF

Internet data arrives at your home through a cable that connects to a modem supplied by your Internet service provider. That modem has an Ethernet port on the back panel. (*Ethernet* is a standard networking technology that supports wired connections between devices.) If you have a computer or other device that also has an Ethernet port or Ethernet adapter, you can connect the device to the modem with an Ethernet cable and enjoy instant Internet access. A direct connection to a modem is the easiest and cheapest way to connect to the Internet, but it's far from the most convenient because:

- You can connect only one device at a time.

- You can't connect devices that don't have a built-in Ethernet port or that can't use an Ethernet adapter.

- You can't connect devices that are too far away from the modem for a direct cable connection.

The networking nerds of the world saw these restrictions and came up with a solution: a router. A *router* connects directly via Ethernet to the modem and then creates a local area network (LAN) to share that connection with other devices in your home.

One way to share that Internet access is by connecting devices directly to the Ethernet ports on the back of the router. But what about devices that don't have Ethernet connectivity or are too far away from the router for a cable connection? That's where wireless connections come in, and for that you need another device called a *wireless access point* (usually shortened to just *access point* or *AP*). With an access point, you can create a wireless network that enables any device (as long the device has the requisite wireless hardware) to connect to the network. What about Internet access? That depends on the device:

- If the access point is a separate device, you connect it via Ethernet to your router. Then any device that connects to the access point's wireless network also gets Internet access via the router.

- If the access point is built into the router (as most access points are these days), devices that connect to the access point's wireless network automatically get Internet access via the router.

Extending Your Wi-Fi Network

The point of a Wi-Fi network is to beam signals from your router to any wireless-capable device in your home (and, of course, to beam data back to the router, as needed). That's the theory, but in practice this simple idea breaks down if you live in a large home or a house that has three or more floors. If you live in such a place, you probably know from bitter experience that your house has at least one — but more likely a few — *dead zones*, where you get a weak Wi-Fi signal at best or no signal at worst.

Want to stream some media to one of these places? Sorry. It's just not going to happen, or the signal will be misery-inducingly bad.

Why do dead zones happen? Two main reasons:

>> The dead zone lies at the edge of or beyond the range of your router's wireless signal.

>> The streaming device's wireless antenna isn't powerful enough to pick up the router's signal.

What can you do about dead zones? Quite a bit, actually:

>> **Move the streaming device closer to the router:** If your streaming device is mobile (such as a smartphone or tablet), try moving closer to the router until you get a stronger signal.

>> **Move the router to a more central location:** In a large or multi-story house, the best location for the router is as close to the geographical center of the house as possible. Moving the router might be a tall order given that it needs a direct Ethernet connection to your Internet modem. In that case, consider asking your Internet provider to move the modem's wall jack to the location you want.

>> **Change the channel:** If your router is broadcasting its signal on a channel that's also being used by your neighbors, you might get degraded performance. You can often fix the problem by changing the router's broadcast to a different channel. See the documentation that came with your router to learn how to choose a different channel.

>> **Add a wireless range extender:** Adding an extender boosts the Wi-Fi signal. Depending on the router, the extender can more than double the normal wireless range, although the resulting signal is usually about half a strong as a full-strength signal.

WARNING

Most range extenders create a separate wireless network, which is a hassle. Look for a range extender that shares the name and password of your existing wireless network.

>> **Set up a wireless mesh network:** A *mesh network* is a wireless network that combines a router and one or more *extension nodes* (also called *satellites*) to extend the full capabilities of the network to every corner of your home. Mesh networks can be expensive but they're usually the most robust way to get full coverage throughout your home.

>> **Upgrade your router:** If you're in the market for a new router, look for a router that has features that extend the wireless range and better enable the signal to penetrate walls and other solid barriers.

How Much Bandwidth Is Enough?

Bandwidth is a measure of how much data gets sent and received along an Internet connection over a specific time frame, usually a month. For example, if you send and receive a gigabyte of data every day for 30 days, your bandwidth for that month is 30 GB. Why is bandwidth relevant to you? For one simple reason:

Streaming media eats bandwidth for lunch. And dinner. Yep, and breakfast, too.

I talk about the specifics of streaming data usage in the next section. For now, the upshot is that when you're reviewing your Internet access, you need to take a hard look at the data portion of your ISP's plans. These plans can be complicated, but they all really boil down to choosing one of the following:

>> **Bandwidth cap:** A maximum amount of bandwidth that you're allowed to use per month. Most big-time ISPs offer a terabyte (1 TB or 1,024 GB) of bandwidth per month. If you go over the cap, you'll be charged a fee, usually around $10 for every 50 GB of extra data. Many ISPs also let you purchase more bandwidth (say, an extra 250 GB for $15 per month).

WARNING

Rather than charge you extra when you exceed your data cap, some ISPs slow down your Internet download speed. That doesn't sound so bad, until you realize that this slower speed might be as low as 1 or 2 Mbps, which is too slow for streaming (see "I Feel the Need — the Need for Speed!" later in this chapter).

>> **No bandwidth cap:** No maximum amount of bandwidth per month. Some ISPs only offer unlimited bandwidth plans. However, with most ISPs, to get unlimited data, you usually have to pay between $30 and $50 extra per month.

How do you choose between these options? If your budget allows, an unlimited data plan is the way to go because then you can stream to your heart's content and never worry about dreaded overage fees (or speed throttling). Otherwise, is a terabyte of data enough? It sure sounds like a lot of data, but remember what I said earlier about the ravenous hunger of streaming media. To know whether a data cap is right for you, you need to know more about how streaming media uses bandwidth, which is what I talk about in the next section.

Taking a look at streaming media bandwidth usage

To make an intelligent guess about how much bandwidth your new streaming habit might use, you need to know the bandwidth rates for the services you use. These rates often aren't easy to come by but are sometimes available in the settings section of the service.

To save you some legwork, the following table lists the approximate bandwidth usage per streaming hour of several popular streaming services.

Service	Standard definition (SD) per hour	High definition (HD) per hour	4K ultra-high definition (UHD) per hour
Amazon Prime Video	600 MB	1.3 GB	5.8 GB
Hulu	675 MB	2.7 GB	7.2 GB
Netflix	700 MB	3 GB	7 GB
Sling TV	500 MB	1.2 GB	2 GB
YouTube	560 MB	3 GB	16 GB

To use Netflix as an example, if you stream 4K video for five hours each day, you'll use 1,050 GB over 30 days, which is just a hair over the 1,024 GB (1 TB) cap that's standard on many Internet plans. By comparison, you could stream about 11 hours of HD video per day and still be under a 1 TB cap.

Going on a bandwidth diet

If you've run the numbers and it looks like your streaming usage is going to consume more than whatever limited data cap you can afford (or if your recent usage history shows that you're coming close to or exceeding your cap), you need to reduce that consumption to avoid extra fees (or a service slowdown).

Fortunately, you can use several techniques to put your bandwidth usage on a stricter diet:

>> **Configure your streaming service to use less data.**
Most streaming services have a setting that lets you choose a stream quality of SD (sometimes shown as low), HD (medium), or UHD (high). Figure 7-2 shows these options in the Playback Settings screen of Netflix.

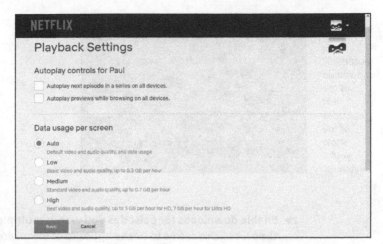

FIGURE 7-2:
Most
streaming
services
let you
choose a
data
usage
setting.

Netflix Playback Settings (as shown in figure):

NETFLIX

Playback Settings

Autoplay controls for Paul

☐ Autoplay next episode in a series on all devices.

☐ Autoplay previews while browsing on all devices.

Data usage per screen

● Auto
Default video and audio quality, and data usage

○ Low
Basic video and audio quality, up to 0.3 GB per hour

○ Medium
Standard video and audio quality, up to 0.7 GB per hour

○ High
Best video and audio quality, up to 3 GB per hour for HD, 7 GB per hour for Ultra HD

[Save]   Cancel

>> **Disable automatic playback.** Most streaming services enable a setting that automatically starts the next episode in a series as soon as you finish watching the current episode. That's convenient, for sure, but it's also an easy way to burn through a bunch of data if you fall asleep (hey, no judgement). Some services (particularly Netflix and Amazon Prime Video) limit the number of episodes that get played automatically, but why waste data if you're bumping up against your data ceiling? Dive into each service's settings and disable the following (the Netflix versions of these settings are shown near the top of Figure 7-2):

- Automatic playback of episodes

- Automatic playback of previews

>> **Disable automatic downloads.** Some services automatically download the next episode of the show you're currently watching. In fact, some services will download the next two, three, or even more episodes! These services are trying to be convenient, but if you don't intend to watch another episode right away or find that you dislike the show you're watching, you've just wasted a ton of bandwidth. Head for the service settings and disable this feature. (Figure 7-3 shows this setting still activated in the Amazon Prime Video mobile app.)

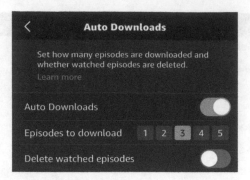

FIGURE 7-3:
Disable
the
automatic
download
of the
next
episodes
of the
show
you're
watching.

>> **Enable downloads for episodes you watch multiple times.** Yes, this seems to be the opposite of the advice in the preceding item, but if you (or, more likely, your kids) will be watching a show multiple times, it's better to download the episode once and then watch it again from your device instead of re-streaming it.

Understanding How Streaming Works

To make the best choices for Internet access, you should understand a bit about how streaming works. As you might imagine, streaming media is a hideously complex bit of business that requires extremely sophisticated hardware and software to make everything work as well as it does. The good news is that you don't need to know anything about that complexity, so you can shut off all those alarm bells ringing in your head. Instead, this section provides you with a basic overview of how streaming performs its magic.

The streaming process

The general process for streaming an on-demand audio or video file is illustrated in Figure 7-4.

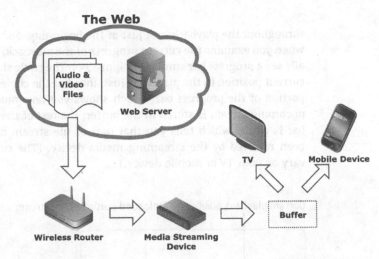

FIGURE 7-4:
An overview of how streaming works.

The Web

Audio & Video Files

Web Server

TV

Mobile Device

Wireless Router

Media Streaming Device

Buffer

As Figure 7-4 shows, streaming is a five-step process:

1. For on-demand audio or video, the media file is stored on the web using a special computer called a *web server*.

2. When a user requests the media, the server begins sending the first few seconds of the audio or video file to the user.

3. When the data reaches the user's network, the network's wireless router passes the data along to the streaming media device.

   Note that the router is usually wireless, but it doesn't have to be.

4. The streaming media device waits until it has a certain amount of the media before it starts the playback.

   The saved data is stored in a special memory location called a *buffer*. (See the next section, "More about buffering," for, well, more about buffering.)

5. When the buffer contains enough data to ensure a smooth playback, the stream is sent to the user's TV or mobile device, and the entertainment begins.

More about buffering

The buffering process that occurs in Steps 4 and 5 in the preceding section is such a crucial part of streaming that it goes on throughout the playback, not just at the beginning. For example, when you examine the current progress of the playback, you usually see a progress bar similar to Figure 7-5. The circle shows your current position in the playback. Just ahead of the circle is a dark portion of the progress bar, which shows you how much of the upcoming stream is stored in the buffer; the rest of the progress bar is white, which tells you that part of the stream hasn't yet been received by the streaming media device. (The colors may vary on your TV or mobile device.)

Current playback position   Unloaded portion of the stream

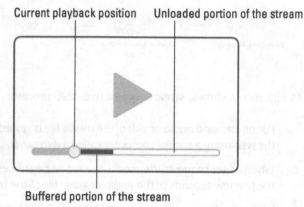

FIGURE 7-5:
Media streams are buffered for smoother playback.

Buffered portion of the stream

Why not just play the media as it arrives and skip the buffer altogether? That would be nice, and it just might work in an ideal world, but the world we inhabit is far from ideal. In real life, media streams can suffer from a number of problems:

» The server may be slow to respond if it has to deal with a large number of media requests.

» Your Internet connection speed may be slow.

» Your network speed may be slow.

» Glitches between the server and your network may mean that large parts of the media stream are delayed or missing.

Any one of these problems could interrupt the stream playback for a split second to a few seconds. Without a buffer to fall back on, your show would have to stop mid-playback to wait for the delay to resolve. However, with anywhere from a few to a few dozen seconds stored in the buffer, the streaming media device can keep the stream playing, and you remain blissfully unaware of any problems because they happen in the background, without affecting your enjoyment of the media.

I Feel the Need — the Need for Speed!

As I explain in a previous section ("How Much Bandwidth Is Enough?"), keeping a metaphorical eye peeled on your bandwidth is crucial, particularly when you're just getting started with this streaming stuff. But your Internet data usage won't amount to very much if your connection is too slow to stream anything. The rest of this chapter takes a close look at one of the most important streaming media metrics: your Internet connection speed.

Why does speed matter?

What's the big deal about Internet speed, anyway? Can't you stream no matter what connection you have? No, I'm afraid not. To understand why, you first need to know that the ravenous appetite that streaming media has for data applies not only cumulatively (for example, how much bandwidth you use in a month) but also in the moment. That is, even a low-quality stream stuffs data through the Internet's tubes at a rate measured in millions of bits every *second!*

That's a torrent of data, and many budget Internet connections just can't handle it. The result? A litany of streaming media problems, including the following:

>> You select Play for some streaming media, but the content never starts.

>> You select Play for some streaming media and the content plays eventually but takes a long time to get there.

>> Streaming media plays for a while, and then stops for a while as it gathers enough data to continue. (This is the buffering process I talked about earlier in the "More about buffering" section.) This play/buffer cycle occurs every few seconds until all your hair is pulled out in frustration.

>> Streaming media plays for a while, perhaps does the play/buffer cycle for a while, and then just stops playing and never resumes.

>> Streaming media plays, but the video or audio or both are often distorted.

>> Streaming media plays, but you get only video, only audio, or the audio and video aren't synced.

Having a too-slow Internet connection is the opposite of fun when it comes to streaming media, so don't try this at home.

How fast is fast enough?

Right, I hear you ask, I get that a too slow connection is bad for streaming, but how fast a connection do I really need? That's an excellent question, and the simplest answer is twofold.

If money is no object (lucky you!), get the fastest connection speed available. If money is very much an object (I feel your pain), get the slowest connection speed that will still enable you to stream.

Okay, I see you scratching your head over that last item, so let's break it down a bit. Basically, you need to select an Internet download speed that takes into account the following:

>> **Stream video quality:** The download speed has to match up with the video quality you'll be using when you stream. That is, you can get away with a slower speed if you'll be streaming everything in SD (480i). But if you want HD (1080p) or even 4K streams, you need to bump up the connection speed accordingly.

>> **Live versus on-demand TV:** You need a bit faster speed for live TV versus on-demand content. (4K live content doesn't exist yet, so the live versus on-demand contrast applies to only SD and HD streams.)

>> **Number of devices:** If multiple devices are accessing the content, you need a faster connection.

>> **Glitch tolerance:** The download speed has to match your own tolerance for stream glitches. That is, there's a bare minimum speed that will get the job done but at the cost of occasional midstream buffering and other problems.

Putting all this together, here's a handy table that shows you, for each stream video quality, the bare minimum speed you need, the acceptable speed for smooth streaming, and the speed required to support multiple streamers.

Video quality	Bare minimum speed	Acceptable speed	Multiple device speed
SD on-demand	3 Mbps	4 Mbps	5 Mbps
SD live	4 Mbps	5 Mbps	6 Mbps
HD on-demand	5 Mbps	8 Mbps	10 Mbps
HD live	7 Mbps	10 Mbps	12 Mbps
4K on-demand	18 Mbps	25 Mbps	40 Mbps

Testing your Internet speed

You might have signed up for an Internet account that claims a particular speed for downloads, but how can you be sure that you're really getting that speed? Fortunately, lots of sites on the web will test your current connection speed:

>> Most ISPs offer a speed test page, so check your ISP's support site for a speed test tool.

>> Run a Google search for *speed test* and then click the blue Run Speed Test button (see Figure 7-6) that appears as part of the first search "result."

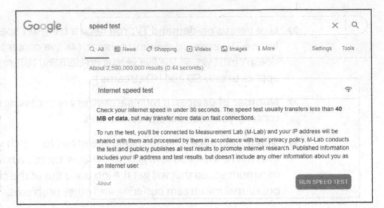

FIGURE 7-6:
Search
Google
for *speed
test* and
then click
the Run
Speed
Test
button.

>> Use any of the following speed test sites:

- Fast.com by Netflix (https://fast.com/)

- Speed of Me (https://speedof.me/)

- Speedtest (https://www.speedtest.net/)

Here are the general steps to follow:

1. **Choose a speed test site.**

2. **Shut down any applications or services that might be downloading or uploading data using your Internet connection.**

3. **(Optional) Reboot both your modem and your router.**

   You don't have to do this if it's a hassle, but you'll get a more accurate result if you reboot both devices.

4. **(Optional) Connect your testing device to your router with an Ethernet cable.**

   Yep, you can run the test over Wi-Fi, but a wired connection to the router is more stable and more accurate.

5. **Run the test.**

   The test usually takes a minute or so. You can see the progress of the test in a window, as shown in Figure 7-7.

FIGURE 7-7: Preliminary results are displayed while the test is running.

Chapter **8**

Checking Out Free Streaming Services

hen you think about streaming media services — you *do* think about them, right? — you probably think about services such as Amazon Prime Video, Disney+, HBO Max, Hulu, or Netflix. That makes sense because every one of those apps is popular and well-known. But do you know what else they all have in common? They all charge a subscription to access their content.

That's reasonable until you get an email informing you that a service has raised its rates *yet again*. Sometimes it's almost enough to make you wistful for the good old days of cable.

Just kidding! But these increasingly frequent rate hikes can make it seem as though paid streaming services view their customers as cash machines from which the services can withdraw money with impunity.

If you're no cash machine or if your streaming budget is already bursting at the seams, you might be interested in an alternative to subscription-based streaming apps: *free* streaming services. That's right: Tons of services offer TV shows, movies, and more for the not-even-close-to-princely sum of zero dollars per month.

In this chapter, you explore the wonderful world of free streaming apps. You learn how they work, what the gotchas are (there are, alas, a few), and what services are available and ready to stream to your favorite device today.

Is There Such a Thing as a Free Lunch?

Having recently freed yourself from the shackles of the cable company, you'd be forgiven for being suspicious of any service or product that describes itself as "free." After all, you're used to the way Big Cable works, which is to charge you an exorbitant monthly fee *and* show you a ton of commercials.

However, trust me when I tell you that there are streaming services out there — quite a few, actually — that won't cost you a dime. Is this particular streaming lunch really free? Yes and no:

>> **No:** Most free streaming services pay the bills by showing you commercials, so the "cost" of the content is that you're subjected to a few ads during each show.

>> **Yes:** A few streaming services are not only free but also don't show commercials. The content is mostly educational, but it's still both cost-free and ad-free. Sweet!

What to Expect from Free Streaming Services

If you've been around the block a time or two, the idea that "You get what you pay for" is probably lodged in your brain (no doubt shoehorned between "A penny saved is a penny earned" and "You can't take it with you"). Yep, it's a cliché, but it's one of those cliches that has the merit of being true. When the cliché is applied to something that's free, it's usually a warning to not expect much from the product or service.

Does "You get what you pay for" apply to free streaming services? Yes and no:

>> **Yes:** Compared to the paid services that I talk about in Chapter 9, free streaming services generally offer less content, fewer features, more restrictions, and lower quality (no 4K streams, in particular).

>> **No:** Most free streaming service run commercials to make ends meet. The good news for you is that in many cases those advertising revenues are substantial enough that the service is able to offer a surprising amount of decent content.

The upshot is that, no, most free streaming services can't offer the bonanza of movies and TV shows that you find with the paid services. But if your budget's tight and your standards are loose, you can easily find a good night's worth of entertainment without having to pay for a streaming subscription.

If you're worried about having to create tons of accounts on these free services, worry no more. Yes, many free streaming services require an account, but a surprising number let you stream as a guest.

Some Free Streaming Services to Check Out

Although the total number of services offering free streaming has fallen from its peak (for example, Hulu stopped offering a free version of its content a few years back), there's still an impressively long list of streaming apps and sites that are gratis. In the sections that follow, I single out what are arguably the ten most popular or useful of these free services, and then I round out the list with quick looks at quite a few other free streaming services (including some for kids).

REMEMBER

If you live outside the United States, many of the services I mention here won't be available to you.

Crackle

www.crackle.com

Crackle is a free, ad-supported service that offers thousands of movies, TV shows, and channels, and even some original content (see Figure 8-1). You can watch Crackle content on its website, on iOS or Android mobile devices, on streaming players from Amazon, Apple, Google, and Roku, and on smart TVs from LG and Samsung.

FIGURE 8-1:
Crackle offers free movies, TV shows, and original content.

Haystack News

www.haystack.tv

If news is your thing, check out Haystack News, an ad-supported streaming service that offers local and world news coverage. Haystack News organizes its content into channels such as Current Events, Science & Technology, Business & Finance, and Sports. The service also offers specific news sources, such as ABC News, CNET, and Al Jazeera News. Choose one or more channels and sources when you sign up and Haystack News customizes your stream accordingly (see Figure 8-2).

FIGURE 8-2: Haystack News offers customized local and global news streams.

You can watch Haystack News on the website, on iOS or Android mobile devices, on streaming players from Amazon, Apple, Google, and Roku, and on most smart TVs, including Android TV.

Hoopla

www.hoopladigital.com

If there's one streaming service that comes pretty darn close to the "free lunch" description, it's Hoopla. Why? Because it's free and it has no advertising. Sweet!

How is it possible for a service to be both subscription- and ad-free? Hoopla's deal is that it offers its content via local public libraries. In the same way that you can borrow, say, a book for free from your local library, you can also "borrow" a movie or television show for free through Hoopla. It's an awesome service, but Hoopla does have a couple of prerequisites:

>> Your local library system must be part of the Hoopla network.

>> You must have a library card in that system.

First, use a computer or mobile device to head over to www. hoopladigital.com and sign up for an account. Part of the sign-up process involves choosing your local library system, as shown in Figure 8-3. (For the easiest selection, allow Hoopla to access your location when prompted by your browser.) You're also asked for your library card number.

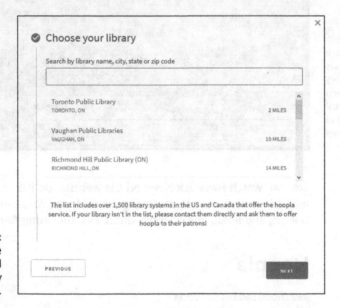

✓ **Choose your library**

Search by library name, city, state or zip code

Toronto Public Library
TORONTO, ON   2 MILES

Vaughan Public Libraries
VAUGHAN, ON   10 MILES

Richmond Hill Public Library (ON)
RICHMOND HILL, ON   14 MILES

The list includes over 1,500 library systems in the US and Canada that offer the hoopla service. If your library isn't in the list, please contact them directly and ask them to offer hoopla to their patrons!

PREVIOUS   NEXT

FIGURE 8-3: Choose your local library system.

After your account is ready, you can access Hoopla content on your computer, on an iOS or Android mobile device, or via the Hoopla app on your streaming player (see Figure 8-4).

FIGURE 8-4:
Use the
Hoopla
app on
your
streamer
to borrow
movies, TV
shows,
and more.

There's got to be a catch, right? Actually, there are two:

>> **There's usually a limit on the number of media titles you can view each month.** That limit depends on the library system.

>> **The selection is often limited.** The offerings vary from one library system to another, but most offer movies and television shows, as well as other digital content, such as music.

IMDb TV

www.imdb.com/tv

IMDb is most famous as the movie and television database site that everyone uses to determine who wins entertainment-related bets (such as whether Dustin Hoffman was really in *Star Wars*). The site also runs a free streaming service called IMDb TV.

The focus at IMDb TV is on older content, which includes thousands of classic movies and TV shows. The service also offers entertainment-related fare such as celebrity profiles, Hollywood documentaries, and award show coverage.

IMDb TV is owned by Amazon, so it's no surprise that you can access the service via Amazon's streaming players by using the IMDb TV app as well as the Amazon Prime Video app. IMDb TV is also available via Roku, Apple TV, and the IMDb TV website.

You can sign up using your Amazon account, or you can create a separate IMDB account.

Kanopy

www.kanopy.com

Kanopy is a service similar to Hoopla, described in an earlier section. That is, Kanopy offers free access to movies, TV shows, and other content without advertising. Like Hoopla, you can access a limited amount of content by using your public library card, but you can also get unlimited access to Kanopy if you're a student or professor at a participating college or university.

PBS Video

www.pbs.org

PBS Video offers a ton of classic PBS content, including Masterpiece, Nova, Nature, and many more (see Figure 8-5). The PBS Video app is available for iOS and Android mobile devices, Amazon, Apple, Google, and Roku streamers and on Samsung smart TVs.

FIGURE 8-5:
PBS Video gives you free access to all your favorite PBS shows.

Peacock

www.peacocktv.com

NBCUniversal's Peacock offers a basic, no-charge, ad-supported version of its subscription service (see Chapter 9 for details). You get access to about two-thirds of the available NBC on-demand shows, plus current TV shows a week after they originally air. You can access shows also from USA Network and Bravo and lots of old and somewhat recent movies.

The Peacock app (see Figure 8-6) is available for iOS and Android mobile devices, Apple, Google, and Roku streamers (but not Amazon Fire TV devices), and on LG and Vizio smart TVs.

FIGURE 8-6: Peacock's free tier gives you lots of content (and lots of ads).

Roku Channel Store

therokuchannel.roku.com

If your home is a Roku shop, you have access to lots of streaming content via the Roku Channel Store (see Figure 8-7). Some of the content requires a paid subscription, but a surprising number of movies and TV shows are free.

The Roku Channel Store is available on all Roku streamers, of course, but there's also an app for iOS and Android mobile devices. If you live in the US, you can also stream Roku Channel content on the Roku website.

FIGURE 8-7:
The Roku
Channel
Store
offers lots
of free
movies
and TV
shows.

Tubi TV

https://tubitv.com/

Tubi is an ad-supported site that offers well over 20,000 titles to stream. These titles include TV shows both old (*The Honeymooners*) and (relatively) new (*The Apprentice*), plus a large selection of movies in every conceivable genre.

The Tubi TV app (see Figure 8-8) is available on a wide variety of devices, including Android and iOS smartphones and tablets, streaming players from Amazon, Apple, Google, and Roku, game consoles such as Xbox and PlayStation, and smart TVs from Samsung, Sony, and Vizio.

FIGURE 8-8:
The Tubi
TV app
offers
free
access to
a large
number
of movies
and TV
shows.

Xumo

www.xumo.tv

Xumo is an ad-supported streaming service that offers more than 180 TV channels that include both live TV feeds and on-demand content. The Xumo app is available on a huge number of devices, including Android and iOS smartphones and tablets, streamers from Amazon and Roku, and a long list of smart TVs, including Android TV, Hisense, LG, Samsung, Sony, and Vizio. You can also stream Xumo content from the company's website.

Free streaming apps for kids

If your parenting strategy includes plopping the kids down in front of a screen while you grab some "me" time, here are a few free streaming apps to check out:

>> HappyKids.tv

>> Kidoodle.TV

>> Kids Movies by Fawesome.tv

>> KidsFlix TV

>> KiddZtube TV

>> Lego TV

>> PBS Kids

>> Pocket.Watch

>> Pocoyo Kids TV

>> Popcornflix Kids

>> YouTube Kids (see Figure 8-9)

Most of these apps are available on iOS and Android mobile devices, as well as on streaming media players from Amazon, Apple, Google, and Roku.

FIGURE 8-9: YouTube Kids offers a ton of kid-friendly streaming content.

REMEMBER

Many of the paid streaming services that I mention in Chapter 9 also offer lots of kid-focused content, so be sure to check out those services if you're looking for ways to keep your kids (and you) entertained.

A few more freebies

For the sake of being semi-complete, here are some brief descriptions of a few more free streaming services to check out in your spare time:

>> **CW Seed:** Lots of old TV shows. Go to www.cwseed.com.

>> **Networks:** The websites and apps of big-time TV networks usually give you at least some free access to shows:

- **ABC:** https://abc.com/
- **BBC:** www.bbc.com
- **CBC:** www.cbc.ca
- **CBS:** www.paramountplus.com
- **The CW:** www.cwtv.com
- **Fox:** www.fox.com
- **NBC:** www.nbc.com

>> **Plex:** Free movies and TV shows. Visit https://www.plex.tv/watch-free.

>> **Pluto TV:** Live TV plus some free on-demand stuff. Go to https://pluto.tv/.

>> **Sling Free:** The free version of Sling TV with a limited selected of on-demand content. Visit www.sling.com/deals/sling-free.

>> **Smart TVs:** Several major brands offer free streaming on their devices:

- **LG Channels:** Streaming content from Xumo and Pluto TV

- **Samsung TV Plus:** Ad-supported news, sports, and entertainment channels

- **Vizio WatchFree:** Streaming content from Pluto TV, plus a collection of ad-supported channels

>> **Vudu:** Movie rentals, but with a section devoted to free movies and TV shows. Go to www.vudu.com.

IN THIS CHAPTER

» **Seeing what you get with a paid streaming subscription**

» **Understanding the different types of paid streaming services**

» **Reviewing a few on-demand services**

» **Eyeballing some cable-replacement services**

» **Taking a look at a few premium channels**

Chapter **9**

Checking Out Paid Streaming Services

O nce you've cut the cord and have washed your hands of the cable company, you're left with a gaping primetime void that needs to be filled with, well, *something*. Huge numbers of cord cutters have cobbled together a satisfying TV watching experience from over-the-air channels (see Chapters 3 through 5), free streaming services (see Chapter 8), and YouTube cat videos. It's entertaining, there's no shortage of content, and it's cheaper than dirt.

The problem with that approach, however, is that you miss out on a ton of the biggest and most popular shows, from *Game of Thrones* and *The Crown* to *Better Call Saul* and *The Marvelous Mrs. Maisel*. These shows and others that have become cultural touchstones are available only on streaming channels and services that require a paid subscription, such as HBO, Netflix, Showtime, and Amazon Prime Video.

Paid services, which are the mainstream (so to speak) of online TV and movie watching, are the subject of this chapter. In the pages that follow, you delve into the riches available via paid streaming apps. You learn how these services work and what types of paid services are available. You also get a long list of the best and most popular apps, what content they offer, how much they cost, what devices they support, and more. It's a veritable streaming feast, so let's dig in.

What to Expect from Paid Streaming Services

One of the nice things about the free streaming apps that I talk about in Chapter 8 is that when you sign up with one of these services, you go in with low expectations. After all, it's free, so how great can it be? If it's not that good, you can say, "See? I told you!" and if it turns out to be pretty decent (as some are) you can say, "See? I told you!"

But when we pay for something, we go in expecting a lot. After all, you're spending your hard-earned money, so the service better deliver, am I right?

Fortunately, most big-time streaming media services offer a decent product for the money. In the end, you'll almost certainly wind up with lots of great content to watch and a monthly bill that's heartwarmingly less than what you were paying for cable.

But life in the fast streaming lane isn't all beer and Skittles. Here are some gotchas to watch out for when you're doing your research:

>> **Commercials:** Some services show ads in exchange for a lower monthly rate.

>> **Availability:** All streaming services support a wide array of platforms on which you can watch their content, including the web, mobile devices, streamers, smart TVs, and gaming consoles. However, no service supports *every* possible

platform, so make sure the service you're interested in is available for your devices.

>> **Add-ons:** The major streaming services offer a huge amount of content, but lots of popular stuff — I'm looking at *you*, HBO Max — is available only as an extra-cost add-on feature.

>> **Simultaneous streams:** You might think that once you've started paying for a streaming service, everyone in your household can now watch content on any device at the same time. This is called *simultaneous streaming*. Most services are cool with it, but usually the number of simultaneous streams depends on the subscription plan. (The more expensive the plan, the more streams you can view at the same time.)

>> **Free trials:** You'd think that every streaming service would offer a free trial of at least a few days so you can test-drive the service before committing. Nope. Most services have a trial period, but a few don't. (I'm looking at *you* again, HBO Max.)

What Types of Paid Streaming Services Are Available?

When deciding where you should spend your no-longer-going-to-the-cable-company money for online entertainment, your first decision involves choosing the type of service. There are three main types:

>> **On demand:** These services offer a library of content that you can demand (that is, watch) any time you feel like it. The main players here are Amazon Prime Video, Hulu, and Netflix.

>> **Cable replacement:** These services aim to provide you with a cable-like experience, which means offering both on-demand TV shows and movies as well as live TV feeds (plus a cloud-based DVR to record live TV). The big-time cable-replacement services are Hulu + Live TV, Sling TV, and YouTube TV.

>> **Premium channels:** These services are mostly focused on providing content from a single channel, although they usually also offer other content. The premium channel services of note are Disney+, HBO Max, and Showtime.

You're not restricted to choosing just one of these categories. For example, many cord cutters choose one on-demand service (such as Netflix or Hulu) and then augment that service with a couple of premium channels (such as Disney+ and HBO Max).

It's a sad fact that many of the services I discuss in this chapter are available only to folks who live in the United States.

REMEMBER

On-Demand Streaming Services

Lots of streaming services provide on-demand TV shows, movies, and other content, but most — the vast majority, in fact — offer a limited number of titles or are brow-furrowingly niche. Awesome as some of these are, I ignore them in this section and instead present the six biggest and most popular on-demand streaming services.

For simplicity throughout this chapter, all subscription prices are listed in U.S. dollars.

REMEMBER

Amazon Prime Video

Website: www.primevideo.com

Free trial: 30 days

Subscription: $8.99 per month or included with Amazon Prime membership ($12.99 per month or $119 per year)

Amazon's Prime Video service (see Figure 9-1) offers unlimited streaming of classic and recent movies and lots of TV shows. Prime Video is also home to some of the best original shows online, including *The Marvelous Mrs. Maisel, Fleabag, Jack Ryan,* and *One Night in Miami.* You can watch streams on up to three devices simultaneously and many shows are streamed in 4K. Are all these features worth $8.99 a month? Almost certainly. However, the big news about Prime Video is that you get everything I just mentioned for no extra cost if you already have an Amazon Prime membership. How awesome is that?

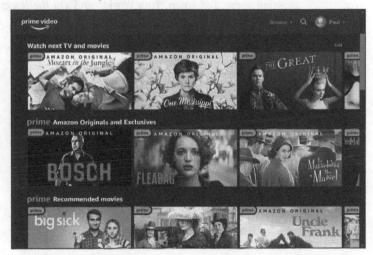

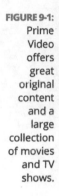

FIGURE 9-1:
Prime Video offers great original content and a large collection of movies and TV shows.

Prime Video also offers rentals and purchases of newly released movies, as well as Amazon Channels, where you can subscribe to big-time streaming channels such as HBO, Showtime, and Starz.

You can watch Prime Video streams either on the web or by using the Prime Video app on Android and iOS mobile devices, on most streaming media players (particular Amazon's own Fire TV streamers), on most smart TVs, and on the Xbox and PlayStation gaming consoles.

Apple TV+

Website: https://tv.apple.com/

Free trial: 7 days

Subscription: $4.99 per month

Apple's on-demand service is unusual in that it offers only original Apple programming (see Figure 9-2). That's right: No classic movies, droll British sitcoms, or kids' favorites here. If that sounds limiting, well, let's just say you're unlikely to subscribe to Apple TV+ as your sole form of streaming content. That said, Apple TV+ is producing some fantastic shows, including *The Morning Show, Ted Lasso, Tiny World,* and *On the Rocks.* You can watch streams on up to six devices simultaneously, and many shows are streamed in 4K.

FIGURE 9-2: Apple TV+ offers only original content on-demand, but some of that content is very good.

Apple TV+ also offers purchases of TV show episodes and seasons, rentals and purchases of new movies, as well as Apple TV Channels, where you can subscribe to major streaming channels such as HBO, Paramount+, Showtime, and Starz.

You can watch Apple TV+ anywhere you can get the Apple TV app, which means iOS mobile devices, Macs, Roku and Fire TV streamers, and many smart TVs.

Hulu

Website: www.hulu.com

Free trial: 30 days

Subscription plans:

Feature	Hulu	Hulu (no ads)
Price per month	$5.99	$11.99
Price per year	$59.99	n/a
Simultaneous screens	2	2
Downloads?	No	Yes
Commercials?	Yes	No
Live TV?	No	No

TIP

Are you a college student? If so, you can subscribe to the ad-supported version of Hulu for a mere $1.99 a month.

Hulu (see Figure 9-3) is one of the top streaming services because it provides a huge amount of content for a decent price (particularly if you don't mind watching commercials). Hulu offers access to most network TV shows as well as some cable channel shows, where in both cases you can see new episodes the day after they originally air. Hulu also produces quite a bit of original content, including *The Handmaid's Tale*, *Little Fires Everywhere*, and *High Fidelity*.

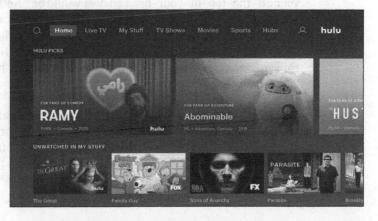

FIGURE 9-3:
Hulu offers great access to network and cable shows, as well as lots of original programming.

Hulu also offers several so-called Premium Add-ons: Cinemax ($9.99 per month), HBO Max ($14.99), Showtime ($10.99), and Starz ($8.99). The service also offers a bundle that includes Disney+ and ESPN+ for $13.99 a month with ads or $19.99 monthly without ads.

You can watch Hulu content either on the web or by using the Hulu app on Android and iOS mobile devices, on most streaming media players, on LG, Samsung, and Vizio smart TVs, and on the Nintendo Switch, Xbox, and PlayStation gaming consoles.

Netflix

Website: www.netflix.com

Free trial: None

Subscription plans:

Feature	Basic	Standard	Premium
Price per month	$8.99	$13.99	$17.99
Simultaneous screens	1	2	4
Mobile devices that can have downloads	1	2	4
Stream resolution	480p	1080p	4K

If someone came up to you and said, "Quick, name a streaming video service!" I'd bet the proverbial dollars to doughnuts that your off-the-top-of-your-head answer would be "Netflix!" (followed no doubt by "Who are you and why are you asking me this?"). Netflix is the quintessential streaming service and the only one with its own pop culture catchphrase ("Netflix and chill"). Even folks who don't know what a streaming media service is know the name *Netflix*.

What has Netflix done to earn its place in our hearts and minds? First, it's been around for a long time, beginning its corporate life as an online DVD-rental outfit way back in 1997 and offering on-demand videos starting in 2007. But over 200 million people rely on Netflix for some or all of their screen-based entertainment because the service has a *ton* of content (see Figure 9-4), including some of the best original shows around, including *The Crown*, *Bridgerton*, *The Queen's Gambit*, and *Stranger Things*.

FIGURE 9-4:
Netflix isn't cheap, but it offers lots of movies, TV, shows, and original content.

Netflix also offers ad-free streaming, offline downloads on mobile devices, and separate user profiles. On the negative side, Netflix is expensive (especially that $17.99 a month for up to four devices and 4K streaming), and it does *not* offer a free trial in most countries.

You can watch Netflix streams either on the web or by using the Netflix app on Android and iOS mobile devices, on most streaming media players, on most smart TVs, and on the Xbox and PlayStation gaming consoles.

Paramount+

Website: www.paramountplus.com

Free trial: One week

Subscription plans:

Feature	Limited commercials	Commercial free
Price per month	$4.99	$9.99
Price per year	$59.99	$99.99
Simultaneous screens	3	3
Mobile downloads?	No	Yes
Live CBS TV feed?	No	Yes
Stream resolution	Some 4K	Some 4K

Paramount+ (formerly known as CBS All Access) offers more than 30,000 TV show episodes and movies from the likes of CBS, BET, Comedy Central, MTV, Nickelodeon, Paramount Pictures, and more. Paramount+ also produces new content, such as *The Good Fight, Star Trek: Discovery,* and the immortal *The SpongeBob Movie: SpongeBob on the Run.* With the Commercial Free plan, you also get live TV shows, news, and sports via your local CBS affiliate.

REMEMBER

The Commercial Free plan should really be named Commercial Free (Sort Of), because you see tons of commercials when you watch the CBS live TV feed.

You can watch Paramount+ content either on the web at www.cbs.com or by using the Paramount+ app on Android and iOS mobile devices, on streaming media players from Amazon and Apple, and on Android TV devices.

Peacock

Website: www.peacocktv.com

Free trial: One week for the Premium plans

Subscription plans:

Feature	Free	Premium (with ads)	Premium (without ads)
Price per month	Free	$4.99	$9.99
Simultaneous screens	3		3
Mobile downloads?	No	No	No
Stream resolution	Some 4K		Some 4K

Peacock (see Figure 9-5) is the streaming service of NBCUniversal (and named after NBC's iconic peacock logo). Peacock Free offers no-charge access to over 13,000 hours of programming, while the two Premium plans bump up the content available to more than 20,000 hours. This content includes classic NBC shows such as *30 Rock, Cheers, Parks and Recreation,* and *The Office.* What? No *Seinfeld*? No *Friends*? Alas not, dear reader. Peacock also has a decent collection of old and recent movies, original shows such as *Brave New World* and *The Capture,* and some live NBC news and sports programs.

FIGURE 9-5: Peacock is NBC Universal's entry into the online streaming game.

You can watch Peacock shows by using the Peacock app on Android and iOS mobile devices, on streaming media players from Apple, Google, and Roku (but not Amazon), on certain smart TVs, and on the Xbox and PlayStation gaming consoles.

Cable-Replacement Services

Some people cut the cable cord but then find that although they're happy to be rid of the cord, they kind of miss the cable. That is, they miss having all their entertainment options — TV shows, movies, live TV — in one convenient place. If that describes you, not to worry: A bunch of online services have as their mission the replacement of cable TV in your home. The next few sections take a look at the four best cable-replacement services.

TIP

Are you in Canada? If so, a worthwhile cable-replacement service to check into is Crave (www.crave.ca).

fuboTV

http://www.fubo.tv/

Free trial: One week for the Family plan only

Subscription plans:

Feature	Family	Elite	Latino Quarterly
Price per month	$64.99	$79.99	$99.99 (three months)
Simultaneous screens	3	8	2
DVR hours	250	1,000	250
Stream resolution	Some 4K	Some 4K	Some 4K

fuboTV offers a mix of on-demand and live channels, more than 160 in all. You get all the US broadcast networks; cable channels such as A&E, Bravo, CNN, and Discovery; and lots of sports, including ESPN, NBA TV, NFL Network, and Golf Channel. fuboTV also offers a cloud-based DVR to record live TV.

If you have the Family plan, you can add the Family Share Max add-on to stream fuboTV on up to five devices on your home network, plus another two devices not on your network. You can add fuboTV's Cloud DVR Plus service, which bumps up the storage hours from 250 to 500. fuboTV also offers several add-on packages that bring you even more channels, including Show-time, Starz, and regional sports coverage.

You can watch fuboTV content by using the service's website or the fuboTV app on Android and iOS mobile devices, on streaming media players from Amazon, Apple, Google, and Roku, on smart TVs from Android, Hisense, and Samsung, and on the Xbox gaming console.

Hulu + Live TV

Website: www.hulu.com/live-tv

Free trial: 7 days

Subscription plans:

Feature	Hulu + Live TV	Hulu + Live TV (no ads)
Price per month	$64.99	$70.99
Simultaneous screens	2	2
Downloads?	No	No
Commercials?	Yes	No
Live TV?	Yes	Yes

Hulu + Live TV offers just what the name says: The full on-demand Hulu streaming service (with or without commercials) that I describe earlier in the chapter (see "Hulu") plus over 60 live channels. The latter include local affiliates of national networks such as ABC, CBS, the CW, Fox, and NBC; news channels such as CNN, Fox News, and MSNBC; and cable channels such as Bravo, E!, TBS, Disney Channel, and National Geographic.

You can also sign up for a couple of add-on features:

>> **Unlimited screens:** Stream to any number of devices connected to your home network, and up to three mobile devices not connected to your network.

>> **Enhanced cloud DVR:** Offers 200 hours of cloud DVR storage and fast-forwarding through ads.

These add-ons are $9.99 per month each, or $14.99 monthly if you get both.

Sling TV

Website: www.sling.com

Free trial: 3 days

Subscription plans:

Feature	Sling Orange	Sling Blue
Price per month	$35	$35
Simultaneous screens	1	3
Channels	45+	30+
DVR storage hours	50	50
Live TV?	Yes	Yes

Sling TV (see Figure 9-6) offers both on-demand content and live TV. The Orange plan is geared towards news (for example, CNN, Fox News, and MSNBC) and entertainment (for example, A&E, AMC, Bravo, and HGTV), while the Blue plan focuses more on sports (for example, ESPN, ESPN2, and ESPN3) and families (for example, Disney Channel, Food Network, and Nick Jr.). A plan that combines both Orange and Blue is available for $50 per month.

FIGURE 9-6: Sling TV replaces your cable service by offering live and on-demand channels.

The basic Sling approach to streaming is to offer a relatively cheap basic plan and then enable you to customize that plan with extras such as DVR Plus (200 hours of storage for $5 per month) and channel bundles such as Sports Extra ($11 monthly), News Extra ($6 monthly), and Kids Extra ($6 monthly). You can also add premium bundles for channels such as ShowTime ($10 per month) and Starz ($9 monthly).

You can watch Sling TV channels by using the Sling TV app on Android and iOS mobile devices, on streaming media players from Amazon, Apple, Google, and Roku, on smart TVs from Android, LG, and Samsung, and on the Xbox gaming console.

YouTube TV

Website: https://tv.youtube.com/

Free trial: 14 days

Subscription: $65 per month

YouTube TV (see Figure 9-7) offers more than 85 channels of on-demand and live TV. These channels include all the major US networks and a good collection of cable content (including AMC, Bravo, Disney Channel, and ESPN). Perhaps best of all, you also get a cloud-based DVR that offers *unlimited* storage. Sweet! You can add premium channels such as HBO Max, Cinemax, Showtime, and Starz for a few extra dollars per month.

FIGURE 9-7: YouTube TV offers lots of channels and unlimited cloud DVR storage.

You can watch YouTube TV on the website or by using the app on Android and iOS mobile devices, on streaming media players from Amazon, Apple, Google, and Roku, on smart TVs from Android, Hisense, LG, Samsung, and Vizio, and on the Xbox and PlayStation 4 gaming consoles.

Checking Out Premium Channels

Most of the on-demand and cable-replacement services that I talk about in this chapter also enable you to add on one or more premium channels. (The adjective *premium* is defined as "Containing one or more shows that you can't live without, therefore we can charge a fortune per month and you'll pay it. Bwaa ha ha ha!") The monthly cost depends on the channel, but it ranges from $6 to a whopping $15 (yes, I'm looking at *you yet again*, HBO Max).

Most streaming devices — particularly Amazon Fire TV, Apple TV, Chromecast, Roku, and most smart TVs — also enable you to subscribe to multiple premium channels from the convenience of your TV.

Table 9-1 lists the six major premium channels, gives you a bit of info on each one, and shows you where to get them.

TABLE 9-1 The Major Premium Streaming Channels

Channel	Standalone Price per Month	Free Trial	Example Content	Also Available On
Cinemax Go www.cinemax.com	$9.99	Depends on how you subscribe, but usually 7 days	*Banshee, The Knick, Outcast*	Amazon Prime Video, Apple TV+, Hulu, YouTube TV, streaming devices
Disney+ www.disneyplus.com	$11.99 ($119.99 per year)	None	*Most Disney movies, Star Wars, Marvel, The Mandalorian, The Simpsons*	Most streaming devices
Epix www.epix.com	$5.99	Depends on how you subscribe, but usually 7 days	*Gemini Man, The Hustle, Pennyworth*	Amazon Prime Video, Apple TV+, fuboTV, Sling TV, YouTube TV, streaming devices
HBO Max www.hbomax.com	$14.99	None	*Game of Thrones, The Sopranos, Friends*	Amazon Prime Video, Apple TV+, Hulu, YouTube TV, streaming devices
Showtime www.sho.com	$10.99	30 days	*Homeland, The Affair, Dexter*	Amazon Prime Video, Apple TV+, fuboTV, Hulu, Sling TV, YouTube TV, streaming devices
Starz www.starz.com	$8.99	Depends on how you subscribe, but usually 7 days	*American Gods, Outlander, Spartacus*	Amazon Prime Video, Apple TV+, fuboTV, Hulu, Roku, Sling TV, YouTube TV, streaming devices

The Part of Tens

IN THIS PART . . .

Getting news, sports, and other streaming services cheaper.

Troubleshooting and solving streaming problems.

Chapter **10**

Ten Ways to Save Money in a Cord-Free World

Y ou might not have many nice things to say about the cable company, but you have to give cable its due when it comes to simplicity: You get a set number of channels for a fixed monthly fee. Done and done. Sure, that so-called fixed fee started high and seemed to get higher all too often, but at least you knew what you were paying month to month.

You can replicate that simplicity when you cut the cord. For example, you can elect to watch just over-the-air TV, which gives you a fixed monthly "fee" of precisely zero dollars. Or you can elect to go with a single subscription, whether it's a streaming service such as Netflix or a cable replacement service such as Hulu + Live TV.

It's more likely, however, that your cord-free experience is going to be more complex and, in general, the greater the complexity, the greater the monthly expense. Although you're likely saving money compared to what you were paying for cable, you might still want to shave a few dollars from your monthly streaming bill.

In this chapter, you discover a fistful of ways to save money in your post-cord life, from making a savvy antenna purchase to eschewing premium Internet and subscription services to not wasting money on unused services. By the end of this chapter, you'll be *making* money. Ka-ching!

Get the Smallest OTA Antenna Possible

If you want to watch over-the-air TV as part of your post-cord lifestyle, who can blame you? You get anywhere from a few to a few dozen channels, all broadcasting in beautifully sharp HD and free for the taking.

Many people are tempted to buy the biggest, baddest antenna on the market and then brag about it on social media. Hey, if that floats your OTA boat, go for it. But if you're interested in saving money (and I know you are or you wouldn't be reading this chapter), you almost certainly don't need the Cadillac of antennas.

Instead, use a service such as TV Fool (www.tvfool.com) to see which over-the-air TV channels are available in your neck of the woods, and then buy an antenna with a range that just exceeds the farthest tower you absolutely must access. Make it an indoor antenna, if possible, because indoor models tend to be less expensive than their outdoor cousins.

Here are a few more ways to save money when putting together an over-the-air TV setup:

>> **Don't bother adding an LTE filter.** You need one of these only if a cellular antenna is right in your neighborhood.

>> **Don't buy multiple antennas.** Remember that a single antenna can provide an OTA signal for multiple devices by using a splitter.

>> **Most folks don't need a signal amplifier.** Note, however, that you might need a distribution amplifier if you want to split your signal and one of those splits has to travel a long way.

>> **Double-check that your antenna comes with its own coaxial cable.** If it does, great: You don't have to buy your own cable.

Don't Get Too Much Internet

As I discuss in Chapter 7, the specs of your Internet connection can make a big difference in your post-cable life. Fundamentally, the quality of your Internet access can turn your cable-free life from nirvana to nightmare if either (or — yikes! — both) of the following are true:

>> **You don't have enough bandwidth.** If your Internet plan has a bandwidth cap, bumping up against that ceiling either means no more streaming TV for you until next month (good thing you've got over-the-air TV, right?) or paying a (usually) obscene rate per gigabyte for going beyond that cap.

>> **You don't have enough download speed.** Streaming video requires a fast Internet connection. (Again, see Chapter 7 to learn how fast.) If your download speeds aren't up to the challenge, your streaming services will either take a very long time to get started or often stop mid-show as your streaming device buffers more content.

So, given those shudder-inducing scenarios, who among us can resist the siren call of unlimited bandwidth and top-of-the-line download speeds? Yes, your streaming problems would be solved, but at what cost? Literally! Internet plans with no bandwidth cap and download speeds measured in the hundreds of megabits per second aren't cheap. Sure, you'll probably still pay less than cable, but with all your other costs for streaming services, your monthly payout could exceed cable without breaking a sweat.

Fortunately, the Internet portion of those costs doesn't have to wreak havoc your budget:

» **Check your bandwidth.** To make an informed decision about how much bandwidth you need, check your usage history. Your Internet provider should have an online tool that shows you how much bandwidth you've used each month.

» **Speed kills (your budget).** If your provider's top-tier Internet plan offers 800 Mbps, 900 Mbps, or even 1 Gbps, forget that. Unless you have people in your house who are serious gamers, these speeds are *way* faster than you need. A plan that offers 50 Mbps or 100 Mbps will get the job done at probably half the cost.

Take Advantage of Skinny Bundles

I'm certain one of the main reasons you bailed on your cable account was because of the dreaded channel bundle. You know what I'm talking about: Your cable company takes one or two super-popular channels — think HBO, Showtime, ESPN — and combines them with a bunch of mediocre and obscure channels. The company then slaps a fat monthly fee on the resulting bundle, which you pay, of course, to get your favorite show. Cue the steam coming out of your ears.

Cutting the cord means you have the option of taking the opposite approach. That is, you can do *a la carte TV*, where you subscribe to only the channels you want to watch. (Examples of a la carte TV services are Apple TV Channels, Amazon Prime Video Channels, and Roku Premium Subscriptions.) This approach gives you more freedom, but those monthly subscription fees can add up to a hefty monthly bill as you add more channels.

Some streaming services offer an in-between option called the *skinny bundle*, which is a relatively small collection of channels for a relatively small monthly fee. Skinny bundles are getting harder to find, but they might be the right option for you if your budget's tight.

Don't Commit Until You're Amazed at Your Luck

The writer Iris Murdock once said that "Writing is like getting married. One should never commit oneself until one is amazed at one's luck." That sage advice also applies to streaming services, believe it or not. Even though so many services are available, many aren't marriage-worthy, as it were. Some are utterly forgettable, most are merely so-so, and a few will seem so right for you that, yes, you'll be amazed at your luck.

Ah, but how will you know which services are dreck and which are gold? The only way to really know is to subscribe and give the service an honest try. Wait a minute, I hear you say, won't that be expensive? Nope, not if you follow these simple steps:

1. Make sure the service offers a free trial for new subscriptions (almost all do).

2. Set up a new subscription for the service.

3. As soon as you get confirmation that your subscription is active, immediately cancel the service. The trick here is that almost all services allow you to complete the free trial, even though you've cancelled your subscription.

4. Check out the service for the duration of the free trial.

5. If you're not amazed by the service, let the free trial period expire and move on with your life. Otherwise, reinstate your subscription, if possible, or start a new subscription and keep it this time.

REMEMBER

Most streaming services make it relatively easy to cancel a subscription, particularly if you subscribed with the service directly. If you subscribed to a service using a streaming media device, such as Roku or Fire TV, you won't be able to unsubscribe using the same device. Instead, you'll need to surf to the service's website and cancel your subscription there.

Watch New Shows on the Cheap

Here's a scenario that's all too familiar to cord-cutters everywhere: A premium channel for which you have no subscription releases a new show that everyone — friends, family, coworkers, people in the off-leash park, total strangers — is talking about. You desperately want to join in the conversation, but it would mean taking out a premium subscription to watch just one show. What to do? Here are some options:

>> **If the service releases an entire season of the show all at once:** Subscribe to the service and then immediately cancel your subscription, as I describe in the previous section. Your free trial should be long enough to watch all the available episodes of the show.

REMEMBER

If you find yourself subscribing and cancelling to the same service multiple times, perhaps it's time to be amazed at your luck (see the preceding section) and get an actual subscription.

>> **If the service releases only individual episodes each week:** Wait until the season is almost over, subscribe, and then cancel the subscription. Again, use the free trial period to catch up with the show.

>> **Wait for the new show to be available on one of your existing services:** Many shows do a first run on one service and are then made available on other services, such as Netflix or Apple TV. If you know the show will be coming to one of your services, just wait for that release. Similarly, networks often stream new episodes for a limited time after the original broadcast, so check the network website for show availability.

Yep, I know: Only the first of these options lets you get in on the current show buzz right away, but that's life in Cord Cutting City.

Subscribe Strategically

One easy way to save money on a streaming service subscription is to pay for an entire year up front, which gets you a discount compared to a month-to-month subscription. (In most cases, you save approximately one or two months' worth of fees.)

However, if you subscribe to a streaming service mostly to watch just one show, you're wasting money by paying for that subscription when no new episodes of your show are available. Saving money in this scenario requires these steps:

1. Switch your streaming service account to a monthly subscription instead of an annual subscription.

2. Watch your favorite show's new episodes.

3. Cancel your subscription.

4. When your favorite show drops new episodes, reinstate your subscription.

5. Repeat Steps 2 through 4 for as long as your show keeps spitting out new seasons.

On a broader level, you can apply a similar strategy to *all* your streaming subscriptions. That is, you subscribe to one streaming service at a time for, say, a month or two. During that time, you watch everything on that service that appeals to you. When the month (or whatever) is up, cancel your subscription, switch to another service, and repeat.

Keep an Eye on Your Subscriptions

Unless you go the one-service-at-a-time route that I describe in the preceding section, your post-cord life will be characterized by having to deal with lots of subscriptions spread across many different services. And unless you have super organization skills, I can guarantee that the more subscriptions you have, the greater the chance that one or both of the following will happen to you:

>> You'll subscribe to a service for the free trial, and then forget to cancel the subscription when the trial period ends.

>> You'll stop watching a service and then forget not only that the service even exists but also that you're still paying for it.

Either way (or both ways), you're wasting money on services you don't use and don't want. Here are some ways to stop the bleeding:

>> If you're just checking out a service, follow the steps I outline earlier in the "Don't Commit Until You're Amazed at Your Luck" section to cancel your subscription immediately after it's activated. That way, if you don't like the service, there's no chance that you'll pay for a few months (or, worse, a whole year) if you forget to cancel.

>> If you are checking out a service but decide not to cancel immediately, use your favorite calendar or task app to create a reminder to cancel the service just before the end of the free trial.

>> Way back in Chapter 2, I suggested making a cord-cutting budget so that you could see if cutting the cord would save you money. If you went to the trouble of making that budget, be sure to update it each time you set up a new subscription. That way, you always have an up-to-date list of all your streaming service subscriptions.

Get a Credit Card Offering Streaming Cash Back

Did you know that you can get paid to watch streaming media? Well, not *paid*, exactly, but you can get some of your streaming fees back. The secret here is to pay for your streaming subscriptions using a credit card that offers cash back for such purchases. Here are some examples:

>> **Amazon Rewards Visa Signature Card:** Offers 5 percent cash back on Amazon Prime Video Channels subscriptions.

>> **American Express Blue Cash Preferred Card:** Offers 6 percent cash back on select streaming subscriptions, including Netflix, Disney+, and HBO Max. Note that this card has a $95 annual fee, so be sure to do the math.

>> **Apple Card:** Offers 3 percent cash back on Apple TV Channels subscriptions and streaming subscriptions set up via iTunes.

>> **US Bank Cash Plus Visa Signature:** Offers 5 percent cash back on select streaming service subscriptions.

TIP

Credit cards aren't the only products that offer streaming goodies. For example, many smartphone plans also include discounts on streaming services.

Give Up the Premium Subscription Plan

Some services offer a deluxe version of their streaming subscription. These premium plans usually come with a few nice perks, but almost always those extras aren't essential for enjoying your shows.

For example, Netflix Premium bumps up the streaming quality to 4K from the standard 1080p. Surely that's a necessity, right, particularly if you have a 4K-friendly TV? Not really. Believe me, shows streamed in good old HD (1080p) look *amazing* on most modern TVs. (The exception is large-screen TVs, which really need 4K; see Chapter 6 for more info.)

And if you do a significant amount of streaming to a mobile device such as a tablet or smartphone, a 4K stream is completely unnecessary. Save your money and ditch the premium plan.

Save on Sports

In Chapter 9, I went through a few options for getting your sports fix via streaming services. The upshot? Streaming sports doesn't come cheap. The sports addict in you might not care, but you should know that there are a few ways to save money when it comes to watching sports. Here are some tips:

» **Go over-the-air.** Fortunately, we still live in a world where the major networks broadcast live sports free over-the-air. Yes, your choices are limited to what's showing live on your local affiliate, but did I mention it's *free*? See Chapter 3, 4, and 5 to get the full scoop on rigging out your entertainment center with over-the-air TV gear.

» **Be strategic.** Earlier (see "Subscribe Strategically") I mentioned that you can subscribe to a service temporarily when it has the content you want. Why not do the same with sports? If you watch only baseball, for example, why pay for an expensive sports package through the winter? If you watch only playoffs for one or more sports, subscribe when the playoffs start, and then cancel your subscription when a winner is crowned.

» **Split the cost with a friend or family member.** Most sport subscription plans enable watching on multiple devices. So get a friend or cousin to go in with you on a single account, and then share the password. Wait a second, isn't that illegal? Not technically, but it's probably not something you should brag about, either.

Chapter **11**

Ten Tips for Troubleshooting Streaming Woes

One of the advantages of having a cable TV account is that if something goes wrong with your connection or hardware, the cable company has an army of technicians at the ready. One call to the company's tech support line and either the technician will be able to solve your problem over the phone or they'll set up a "truck roll," where a technician comes to your residence to make the repair. Now, of course, we're talking about the cable company here, so that appointment probably is far in the future, will require you to take time off work so that you're around when the technician shows up, and will cost you big bucks.

When you cut the cord and leave the cable company behind, you also leave behind the expertise of the cable technician, which means if something goes wrong with your streaming, you're on your own when it comes to fixing things.

That situation is not as daunting as it might sound because most streaming glitches can be solved fairly straightforwardly, even if you have zero technical skills. In this chapter, you learn all the know-how you need to troubleshoot and solve the most common media streaming woes. Whether it's excessive buffering, slow connections, or blank screens, this chapter shows you how to investigate the problem and how to solve it.

Check Your Streaming Device's Download Speed

Here are the most common (and most frustrating) complaints I hear from streamers:

>> The media never starts.

>> The media takes a long time to start.

>> The media plays intermittently.

>> The media stops playing and never resumes.

It's maddening, for sure, but most of the time you can fix the problem. I say "most of the time" because there are a couple of situations where media streaming just doesn't work well:

>> **When you have a slow Internet connection speed:** Media files are usually quite large, so for these files to play properly you need a reasonably fast Internet connection. Amazon recommends at least a half a megabit per second (0.5 Mbps), but realistically you should probably have a connection that offers at least 8 Mbps download speeds for HD content. (See Chapter 7 for more details.)

>> **When you have an intermittent Internet connection:** If you live in an area with spotty Internet service, that now-you-see-it-now-you-don't Internet connection makes streaming media impossible.

If you live in an area that's supposed to have zippy Internet download speeds, the preceding problems can result from a slow connection. I talk about how to check the speed of your Internet connection in Chapter 7.

However, even if your Internet download speeds are good, that only gets you as far as your router. You might be having streaming glitches due to a connection problem *after* the signal leaves the router. There are two things to consider here:

>> **Wi-Fi connection speed:** The problem might be caused by a too slow Wi-Fi network connection. I cover this possibility a bit later in the "Check Your Wi-Fi Connection Speed" section.

>> **Streaming device connection speed:** The problem might be caused by a snafu in the streaming device. I cover this possibility in the rest of this section.

Your streaming device might have some sort of malfunction causing it to process streams at a much slower rate than what you'd normally expect given your Internet speed and Wi-Fi setup. How can you tell? It depends on the device, but most streaming devices offer a way to check the device's current Internet download speed:

>> **Setting:** See if your device has a setting that enables you to check the device's download speed. On Roku devices, for example, select Network, select Settings, and then run the Check Connection command.

>> **App:** Most streaming devices offer multiple apps that can check both download and upload speeds. (Figure 11-1 shows an example.) Search your device's apps for *speed test* and then install and run one of them.

FIGURE 11-1:
Most streamers offer apps that can test your device's Internet speeds.

>> **Web browser:** If your streaming device offers one or more web browsers, you can install a browser and then use any of the Internet speed websites that I mention in Chapter 7. For example, if you have Fire TV, you can use either the Amazon Silk web browser (see Figure 11-2) or the Firefox web browser.

FIGURE 11-2:
On Fire TV, use the Amazon Silk browser to access Internet speed tests.

TIP

If your Internet connection speed seems normal but your streaming device says the download speed is molasses-in-January slow, it's possible that the device itself is the problem. You might be able to get the device back up to speed by doing the following three things (each of which I explain in more detail in the sections that follow):

>> Restart your streaming device.

>> Update your streaming device's system software.

>> Reset your streaming device to its factory default settings.

REMEMBER Try restarting your streaming device to see if it solves your problem. If not, move on to updating the software and see if that helps. If there's still no joy, only then should you try resetting your device to its factory default settings.

Restart Your Streaming Device

If your streaming device is having trouble playing media, connecting to Wi-Fi, pairing with a Bluetooth device, or doing any of its normal duties, by far the most common solution is to shut down the device and restart it. By rebooting the device, you reload the system, which is often enough to solve many problems.

Use either of the following techniques to restart your streaming device:

>> If you still have access to the device interface, access the device settings and run the Restart command.

>> If the device uses electrical power from an outlet, unplug the device's power cord, and then plug it back in.

WARNING You may be tempted to just plug the streaming device back in again right away but hold on a bit. The device has internal electronic components that take some time to completely discharge. To ensure that you get a proper restart, wait at least three seconds before reconnecting your streaming device's power supply.

Check Your Streaming Device for Updates

Your streaming device uses internal software to perform all sorts of tasks, including connecting to your Wi-Fi network, handling media playback, and saving your settings. If your device is acting weird and restarting the device doesn't help, you can often un-weird the device by updating its internal software. Sometimes installing a new version of the operating system is all you need to make the problem go away. In other cases, updating the system may fix a software glitch that was causing your problem.

Here are the general steps to follow to check for and install software updates on most streaming devices:

1. **Open the device settings.**

2. **Locate and choose the command that checks to see if any updates are available.**

   If an update is waiting, the device usually downloads the update and then displays a command for installing the update, as shown in Figure 11-3.

3. **Choose the command to install the update.**

   The streaming device installs the update. During this process, leave your device on and don't press any buttons on the remote.

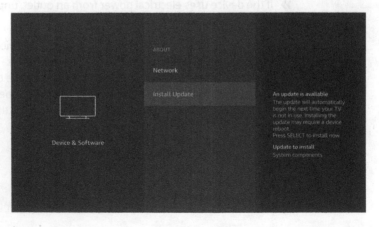

FIGURE 11-3: The Install Update command appears when Fire TV has downloaded an update to Fire OS.

Reset Your Streaming Device

If your problem is particularly ornery, restarting or updating the device won't solve it. In that case, you need to take the relatively drastic step of resetting your streaming device to its original (often called *factory default* or just *factory*) settings. I describe this step as drastic because it means you have to go through the device setup process all over again, so only head down this road if restarting and updating your device don't solve the problem.

Before resetting your streaming device, you might want to check your Wi-Fi connection speed, as I describe in the next section. If your Wi-Fi is operating normally, *then* reset your device.

Here are the general steps to follow to reset your streaming device:

1. **Open the device settings.**

2. **Locate and choose the command that resets the device.**

   Figure 11-4 shows an example. The streaming device resets and then restarts.

FIGURE 11-4: Look for the command that resets the device to its original settings.

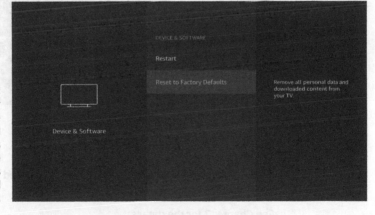

Check Your Wi-Fi Connection Speed

Okay, your streaming is still slow and choppy. You've checked your Internet connection speed, and it's solid. You've restarted, updated, and reset your streaming device, so it's good to go. What's next? The last link in the chain is your Wi-Fi network's connection speed. If the streaming data is arriving at your modem lickety-split and your streaming device is firing on all cylinders, none of that matters if your Wi-Fi router is beaming the data to the device at a snail's pace.

First, check your network signal strength using your device's settings. Figure 11-5 shows an example. If the signal strength is low or fair, see the next couple of sections for some remedies.

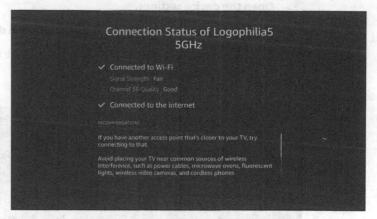

FIGURE 11-5: Many streaming devices enable you to see the Wi-Fi network signal strength.

Here's another way to gauge Wi-Fi connection speed:

1. **Determine your current Internet download speed using a device connected directly to your modem.**

   See Chapter 7 for the details.

2. **Bring a mobile device or notebook computer near the streaming device and then run an Internet speed test on the mobile device or computer.**

3. **Compare the results of Steps 1 and 2 with the following table.**

Wi-Fi standard	Theoretical speed	Real-world speed
802.11b	11 Mbps	2-3 Mbps
802.11a	54 Mbps	20 Mbps
802.11g	54 Mbps	20 Mbps
802.11n	600 Mbps	40-50 Mbps
802.11ac	1.75 Gbps	100 Mbps

TIP

How do you know which Wi-Fi standard your network uses? The most reliable way to tell is to access your Wi-Fi router configuration page, which should tell you which mode your network uses. See your router's documentation to learn how to access the configuration page.

What are you looking for here? There are two scenarios to consider:

>> **Your Wi-Fi network's real-world speed is the same as or faster than your Internet download speed.** In this case, you have a Wi-Fi problem if the mobile device or computer speed test from Step 2 returns a download speed that's significantly less than the Internet speed from Step 1.

>> **Your Wi-Fi network's real-world speed is less than your Internet download speed.** In this case, you have a Wi-Fi problem if the mobile device or computer speed test from Step 2 returns a download speed that's significantly less than your Wi-Fi network's real-world speed.

If the data show you have a Wi-Fi problem, see the next couple of sections for some solutions.

TIP

I should also mention a third possibility here. Move your mobile device or computer closer to your Wi-Fi router and run the Internet speed test again. If you see greatly increased speed compared to what you saw near your streaming device, you have a range problem. That is, your streaming device is too far from your router.

Reset Your Wi-Fi

If your Wi-Fi network isn't performing as it should, try these remedies, in the order shown:

» **Restart your Wi-Fi hardware.** Reset your hardware by performing the following tasks, in order:

1. *Turn off your modem.*

2. *Turn off your Wi-Fi router.*

3. *After a few seconds, turn the modem back on and wait until the modem reconnects to the Internet, which may take a few minutes.*

4. *Turn on your Wi-Fi router.*

REMEMBER

Many Wi-Fi devices these days are all-in-one gadgets that combine both a Wi-Fi router and a modem for Internet access. If that's what you have, instead of performing Steps 1 through 4, you can just turn off the Wi-Fi device, wait a bit, turn the device back on, and then wait for the device to connect to your Internet service provider (ISP).

» **Update the wireless router firmware.** The wireless router firmware is the internal program that the router uses to perform its various chores. Wireless router manufacturers frequently update their firmware to fix bugs, so you should check whether an updated version of the firmware is available. See your device documentation to find out how this works.

» **Reset the Wi-Fi device.** As a last resort, reset the Wi-Fi router to its default factory settings. (See the device documentation to find out how to do this.) Note that if you do this, you need to set up your network again from scratch.

Make Some Wi-Fi Adjustments

Here are a few troubleshooting tips to try if your Wi-Fi performance isn't what you expect:

>> **Shut down other wireless devices.** If you have other devices accessing your Wi-Fi network, shut down any devices you're not using.

>> **Look for interference.** Devices such as baby monitors and cordless phones that use the 2.4 GHz radio frequency (RF) band can wreak havoc with wireless signals. Try either moving or turning off such devices if they're near your Fire TV device or Wi-Fi device.

WARNING

TIP

Keep your Wi-Fi router well away from microwave ovens, which can jam wireless signals.

Many wireless routers enable you to set up a separate Wi-Fi network on the 5 GHz RF band. This band isn't used by most household gadgets, so it has less interference. Check your router manual to see if it supports 5 GHz networks.

>> **Check your range.** Your streaming device may be too far away from the Wi-Fi router. You usually can't get much farther than about 230 feet away from most modern Wi-Fi devices before the signal begins to degrade. (That range drops to about 115 feet for older Wi-Fi devices.) Either move the streaming device closer to the Wi-Fi router or turn on the router's range booster, if it has one. You could also install a wireless range extender.

Troubleshoot an Unresponsive Streaming Device

Perhaps the most teeth-gnashingly frustrating problem you can encounter in technology is when a device — particularly one you paid good money for — just stops working. The device appears to be on, but tapping it, shoving it, gesticulating at it, and yelling at it are all ineffective at making the device respond.

If that happens to your streaming device, try the following troubleshooting steps, in order:

1. Wait a few minutes.

Sometimes devices just freeze up temporarily and then right themselves after a short break.

2. Check your Wi-Fi network to make sure it's working properly and that your device is connected.

See the previous three sections in this chapter.

3. Restart the device.

See the "Restart Your Streaming Device" section, earlier in the chapter.

4. Check to see if your device is using the most up-to-date system software.

See the "Check Your Streaming Device for Updates" section.

5. Reset your device.

See the "Reset Your Streaming Device" section.

Troubleshoot a Blank TV Screen

If the TV to which you've connected your streaming device shows a blank screen, here are a few things to check out:

» Make sure the TV is plugged in and turned on.

» Make sure the TV is set to the correct input:

  • For a TV with a streaming device attached, switch to whatever input the streaming device is connected to.

  • For a Smart TV, make sure the TV is using the streaming input (and not, say, the input for your antenna or gaming console).

» For a streaming device connected to the TV via HDMI, disconnect and then reconnect the device.

>> If your streaming device is connected to your TV with an HDMI cable or HDMI hub, try replacing the cable or the hub or both.

>> If you have a streaming device that supports 4K, make sure you're using a high-speed HDMI cable.

Upgrade Your Hardware

All the troubleshooting tips so far in this chapter haven't cost you a dime, which is the best kind of tech first aid. However, things break or degrade, so if you still have streaming woes, despite your best efforts, new hardware might be in order:

>> If you've tried everything to improve Wi-Fi speeds and nothing works, it's probably time for a new router.

>> If Wi-Fi is fast near the router but not by your streaming device, and you can't move either device so they're closer, you might want to invest in a Wi-Fi range extender.

>> If you have a fast Internet connection and a fast Wi-Fi connection and if you've restarted, updated, and reset your streaming device but that device still has trouble streaming, consider investing in a new streaming device.

Glossary

1080p: A screen resolution that uses 1,080 horizontal lines with progressive scanning. *See also* HDTV.

4K: *See* ultra HD.

480i: A screen resolution that uses 480 horizontal lines with interlaced scanning. *See also* SDTV.

576i: A screen resolution that uses 576 horizontal lines with Interlaced scanning. *See also* SDTV.

720p: A screen resolution that uses 720 horizontal lines with progressive scanning. *See also* HDTV.

8K: *See* ultra HD-2.

a la carte: A streaming media service where you pay for just the channels you want to watch.

access point: *See* wireless access point.

aerial: *See* antenna.

AirPlay: A technology found in some streaming media players that enables you to beam media from an iPhone, iPad, or Mac directly to the player.

amplifier: *See* signal amplifier.

antenna: A device that can receive over-the-air TV signals broadcast from a transmission tower that's within the antenna's range.

antenna combiner: *See* antenna coupler.

antenna coupler: A device that combines the signals from multiple antennas into a single output.

antenna gain: The number of decibels (db) of power that an antenna's amplifier boosts the incoming signal.

ATSC 3.0: In the United States and Canada, a new broadcast TV standard that offers higher picture quality, improved signal reception (especially for indoor antennas), interactive features, and targeted advertising.

band: A radio frequency over which a device sends and receives data.

bandwidth: A measure of how much data gets sent and received through an Internet connection during a specified time frame, such as a month.

bandwidth cap: A maximum amount of bandwidth an Internet connection is allowed to use during a specified time frame.

broadcast TV: *See* over-the-air.

buffer: An area of memory or storage that's used to store the next few seconds or minutes of streaming playback.

cable TV: A television signal carried via a cable.

cable-replacement service: A streaming service that offers TV show and movie bundles similar to those offered by cable services.

channel bundle: A high-priced cable company offering that combines one or two premium or popular channels with a bunch of second-rate channels.

closed captioning: *See* subtitles.

cloud DVR: A DVR that stores its recordings online.

coaxial cable: A cable used to connect an over-the-air antenna to a TV, an external tuner, or a DVR.

coaxial cable splitter: A device that takes a single coaxial cable connection and offers two or more coaxial outputs so that you can split an over-the-air TV signal between multiple devices.

coaxial extension adapter: A device that enables you to connect two coaxial cables.

content: Another term for *media.*

cord avoider: A person who looks for online alternatives to paying for cable TV offerings.

cord cutter: A person who severs her relationship with her cable company and finds alternatives to cable elsewhere.

cord hater: A person who really dislikes paying for cable TV.

cord never: A person who has never had a cable TV account.

cord shaver: A person who takes steps to reduce her cable TV bill.

cord trimmer: *See* cord shaver.

customer retention agent: A cable company employee whose job it is to convince people like you not to cancel their accounts.

data cap: *See* bandwidth cap.

dead zone: An area of your home that gets a very weak Wi-Fi signal or no signal.

digital cliff: A phenomenon that causes a weak over-the-air TV signal to not appear on the TV (as opposed to appearing on the TV with a poor or intermittent picture).

digital video recorder: *See* DVR.

directional: *See* unidirectional.

discoverable: A Bluetooth device that's broadcasting its availability for pairing with another device.

distribution amplifier: A coaxial splitter that also amplifies the signal to make up for the power loss that occurs when you split a signal.

dongle: A device that plugs directly (that is, without a cable) into a port on another device, such as a computer or TV; *see also* streaming dongle.

download speed: The rate — usually measured in Mbps — at which data is sent from the Internet to your location; *see also* upload speed.

dual band: An over-the-air TV antenna that supports both VHF and UHF frequencies.

DVR: A device that records a live TV signal for later playback.

Ethernet: A networking technology that enables wired communications between devices.

extension adapter: *See* coaxial extension adapter.

extension node: A device that helps extend a mesh network.

F connector: The endpoint of a coaxial connection, which will be a plug (usually at both ends of a coaxial cable) or a jack (usually part of a device, such as a TV or DVR).

F-type connector: *See* F connector.

factory default: The original settings that are the defaults for a device when it's new. Resetting a device to its factory default settings can often solve recalcitrant problems.

free trial: A period — usually a month — during which you can try a streaming service without charge.

full-power station: A local TV station that uses a transmission tower with a range of between 50 miles (80 kilometers) and 80 miles (128 kilometers).

gain: *See* antenna gain.

Gbps: Gigabits per second; a measure of the speed of a data transmission. *See also* Mbps.

GHz: Gigahertz; a measure of electromagnetic wave frequency.

gigabit: A billion bits.

gigahertz: A billion hertz.

HD: *See* HDTV.

HDTV: A screen resolution that supports both 1080p and 720p.

HDTV antenna: An antenna that can receive over-the-air HDTV signals.

HDTV tuner: A tuner that can interpret over-the-air HDTV signals.

hertz: An electromagnetic wave frequency equal to one cycle per second.

hi-V: *See* high-VHF.

high-VHF: In the United States and Canada, the VHF range for TV transmissions between 174 MHz and 210 MHz for channels 7 through 13; *see also* low-VHF.

horizontal resolution: The number of pixels in a single scan line from left to right across a TV screen.

Hz: Hertz; a measure of electromagnetic wave frequency.

independent station: A TV station that is not affiliated with any national broadcast network and is independently owned and operated.

input source: The incoming connection that the TV uses to display a signal on its screen.

intelligent band steering: Enables a Wi-Fi router to automatically choose the best band available for the data it's receiving.

interlaced scanning: A TV technology in which the odd and even scan lines of the screen are rendered separately.

jack: A connector into which you can insert a plug; *see also* F connector.

Internet TV: A television signal carried via a video stream over the Internet.

line-of-sight: An imaginary straight line between an antenna and a transmission tower that doesn't go through the Earth or through a tall object such as a building or a hill.

linear TV: *See* over-the-air.

livestreaming: As-it's-happening audio or video, such as on-the-air TV programs delivered by your cable provider or Fire TV Recast, live concerts or sporting events, Internet-based audio or video phone calls, or video feeds of a specific place or scene.

lo-V: *See* low-VHF.

low perceived value: A product or service for which customers don't feel like they're getting their money's worth; also known as cable TV.

low-power station: A local TV station that uses a transmission tower with a range between 15 miles (24 kilometers) and 30 miles (48 kilometers).

low-VHF: In the United States and Canada, the VHF range for TV transmissions between 54 and 82 MHz for channels 2 through 6; *see also* high-VHF.

LTE filter: A device that improves over-the-air TV signal quality by filtering out interference from surrounding LTE cellular signals.

Mbps: Megabits per second; a measure of the speed of a data transmission. *See also* Gbps.

media: Anything you can play via Fire TV, including movies, TV shows, games, music, slideshows, and home videos.

megabit: A million bits.

member station: A TV station that is part of a collection of stations that together own the network. The main example in the US is the Public Broadcasting Service (PBS). Each member station is independently owned and operated.

mesh network: A wireless network that combines a router and one or more extension nodes. These devices work together to extend the full capabilities of the network to every corner of your home.

modem: A device that receives data from and sends data to the Internet.

multidirectional: An antenna that picks up signals from any direction; *see also* unidirectional.

multipath distortion: Over-the-air TV signal interference caused by a single broadcast signal bouncing off buildings and other objects, resulting in multiple reflections of the signal reaching an antenna.

network affiliate: A TV station that carries some or all programs broadcast by a particular national broadcast network, but the station is independently owned and operated.

network feed: Television programming supplied to a local station by a parent national broadcast network.

Next Gen TV: *See* ATSC 3.0.

O&O: *See* owned and operated.

omnidirectional: *See* multidirectional.

OTA: *See* over-the-air.

OTT: *See* over-the-top.

over-the-air: A live TV broadcast signal that can be picked up by using an HDTV antenna.

over-the-top: A TV service delivered via a broadband Internet connection.

overdriving: Making an over-the-air TV signal too powerful for a TV tuner to pick up by using a signal amplifier to boost an already strong over-the-air TV signal.

owned and operated: A TV station that is the property of, and is run by, a national broadcast network (such as one of the so-called Big Five in the United States: ABC, CBS, the CW, Fox, or NBC).

pair: To connect two Bluetooth devices.

pass-through fees: Government-mandated regulatory fees that the cable company is all too happy to pass along to its customers.

pixel: A picture element; a point of light that uses a combination of red, green, and blue.

plug: A connector that you can insert into a jack; *see also* F connector.

power over coax: A device that supplies electrical power via a coaxial cable.

progressive scanning: A TV technology in which all the scan lines of the screen are rendered at the same time.

quad shield: A coaxial cable that offers four layers of shielding to help prevent signal leakage; *see also* triple shield.

range: The maximum distance that a transmission tower can be located from an antenna for the antenna to pick up the tower's broadcast signal.

range extender: *See* wireless range extender.

resolution: The sharpness of a TV screen, measured in pixels.

RF channel: *See* transmit channel.

router: A device that connects directly via Ethernet to an Internet modem and then creates a local area network to share that connection with other devices in your home. *See also* wireless access point.

satellite: *See* extension node.

satellite TV: A television signal transmitted via an orbiting satellite.

SD: *See* SDTV.

SDTV: A screen resolution that supports both 480i and 576i.

second screen experience: Watching media on a main screen (such as your TV) and using a second screen (such as tablet or smartphone) to control playback and display extra info about the media, such as the cast and music.

self-oscillation: *See* overdriving.

set-top box: A box-like streaming media player that's meant to sit on a shelf or table and connect to your display device using an HDMI cable; *see also* dongle.

signal amplifier: A device that boosts an over-the-air TV signal.

signal attenuator: A device that reduces an over-the-air TV signal.

signal splitter: *See* coaxial cable splitter.

simultaneous streaming: A streaming media feature that enables you to access the service's content on multiple devices at the same time.

skinny bundle: A streaming media service bundle that includes only a small number of channels.

smart TV: A TV that has computer hardware on the inside that runs essentially the same software as a streaming media player.

splitter: *See* coaxial cable splitter.

streamer: *See* streaming media player.

streaming: A method of sending media over the Internet from a server to a computer in which the media begins playing within a few seconds and continues until the media is complete.

streaming dongle: A streaming media player that connects directly to an HDMI port on a TV or display.

streaming media: Television programs — as well as movies, music, podcasts, and other media — that are made available over the Internet.

streaming media player: A device that enables you to access and watch streaming media.

streaming media playback: The capability of playing, pausing, rewinding, and fast-forwarding an incoming media stream, usually by pressing buttons on a remote control that comes with the streaming media device.

streaming service interface: A method for discovering and interacting with services that offer audio, video, or live streams.

streaming soundbar: A device that combines a streaming media player with audio hardware, usually including speakers and subwoofer.

streaming stick: *See* streaming dongle.

subtitles: Text transcriptions of the voice track in a TV show, movie, or video.

terminator: A device placed over an unused coaxial cable output to prevent signal leakage from that output.

terrestrial TV: *See* over-the-air.

transmission tower: A tall structure that uses a transmitter to broadcast a station's digital television signal as radio waves in all directions.

transmit channel: The channel number corresponding to the frequency at which an over-the-air TV signal is broadcast; *see also* virtual channel.

tri-band: A router feature that offers three radio bands.

triple shield: A coaxial cable that offers three layers of shielding to help prevent signal leakage; *see also* quad shield.

tuner: A device that can interpret the signals received from an over-the-air antenna and display those signals as television video and audio.

UHD: *See* ultra HD.

UHD-2: *See* ultra HD-2.

ultra HD: A screen resolution of 3840 x 2160 pixels.

ultra HD-2: A screen resolution of 7680 x 4320 pixels.

ultra-high frequency: *See* UHF.

UHF: Radio frequencies ranging from 300 MHz and 3 GHz. The specific range used by TV signals varies by country.

unidirectional: An antenna that picks up signals from a single direction; *see also* multidirectional.

upload speed: The rate — usually measured Mbps — at which data is sent from your location to the Internet; *see also* download speed.

variable amplifier: A signal amplifier that offers an adjustable rate of amplification, usually to prevent overdriving the signal.

vertical resolution: The number of pixels in a vertical column from top to bottom of a TV screen.

very high frequency: *See* VHF.

VHF: Radio frequencies ranging from 30 to 300 MHz. The specific ranges used by TV signals vary depending on the country.

video streaming: Mostly prerecorded TV shows and movies through services such as Amazon Prime Video and Netflix.

virtual channel: The channel number you tune to on your TV to watch an over-the-air TV channel, which is usually different than the station's transmit channel.

wireless access point: A device that creates a wireless network. Most wireless access points today are bundled with a router.

wireless range extender: A device that increases the range of a wireless signal.

Index

About the Author

Paul McFedries has been using, programming, building, yelling at, and occasionally physically attacking computers since he was in short pants (that's a long time). Paul is the author of 100 books that have sold more than 4 million copies throughout the solar system. Among his recent Wiley titles are *Amazon Fire TV For Dummies*, *G Suite For Dummies*, and *iPhone Portable Genius*. Paul invites everyone — especially *you!* — to drop by his personal website (www.mcfedries.com) or follow him on Twitter (@paulmcf).

Dedication

To Karen and Chase, who make life fun.

Author's Acknowledgments

Master of mysteries and lord of law...the editor spills his [or her] will along the paper and cuts it off in lengths to suit.

—Ambrose Bierce

Do you ever think about another person's job and say to yourself, "Man, how do they do that?" That happens to me all the time when I think about editing, because an editor is, as Ambrose Bierce says, "master of mysteries and lord of law." The finer points of sentence structure and production codes are opaque to most of us, but they're an editor's bread and butter and paying attention to them always results in better and clearer prose. So, I'm a big fan of editors because they make me look good! You should be a fan, too, because the editors who worked on this book made it a much better read.

I offer my sincere gratitude to everyone at Wiley who made this book possible, but I'd like to extend a special "Thanks a bunch!" to the folks I worked with directly: executive editor Steven Hayes, project editor and copy editor Susan Pink, and technical reviewer Ryan Williams.

Publisher's Acknowledgments

Executive Editor: Steve Hayes

Project Editor: Susan Pink

Copy Editor: Susan Pink

Technical Editor: Ryan Williams

Proofreader: Debbye Butler

Production Editor: Mohammed Zafar Ali

Cover Image: © Andrey_Popov/ Shutterstock